Praise for Bill Irving

"Bill Irving connects weather data over time to the real stories of actual people we know in Coeur d'Alene to give a true and human picture of what climatic warming has done, is doing now and could do in the future, right here in the Inland Northwest. A counselor by trade, Bill tells not just what climate change is doing to the weather, but to the human heart."

Mike Bullard, author,
Lioness of Idaho: Louise Shadduck and the Power of Polite

"Every home in North Idaho should have a copy of Bill Irving's well-researched book on climate changes that are taking place right here before our non-aware eyes. His important book is a warning of what future changes lie ahead in our North Idaho climate. Our communities need to anticipate and plan. Bill Irving's book is an eye-opener for individuals and families on the importance of changing our energy use habits."

Mary Lou Reed,
Idaho Senator, 1984-1996,
founder, Human Rights Education Institute

"Recently, I had the privilege of reviewing Bill Irving's "Losing Home: Grief and Hope In A Changing World." This is a rigorously researched work surrounding our local Kootenai County, and the very real impact of climate change - socially, politically and economically. If you believe that our county is not affected or minimally so by climate change, and more specifically the warming of our local climate, you must read this book. Mr. Irving has spared no effort in documenting the temperature changes over decades, the impact to our environment, and the likely impact if regional warming continues unabated. There is no doubt after reading this manuscript, that if we disagree with the warnings in this book, I believe that we do so at our own peril. This is not an alarmist work, though a work designed to heighten our awareness - scientifically - to the likely impacts of continued warming, and especially to our local "gem" as we refer to it, Coeur d'Alene Lake. We've already witnessed the change from decades ago when locals would cross the frozen Lake, typically to usher their young ones to school. We've also witnessed the impact to the lake as oxygen levels decrease with increased warming (among other causes), thereby threatening the health of the various species that either occupy or feed from the Lake. We've also witnessed the challenges to our human population, as those suffering from asthma, and the elderly, are at times forced to shelter in cooling stations around the county if they're not equipped with air conditioning. Incidents of respiratory emergent conditions have also taxed our local medical resources, which until recently further exasperated our overworked medical professionals, battling the COVID pandemic. Please, I implore you, and especially if you are a public elected official, read this book, an essential work for public policy formulation."

Chris Fillios,
former Kootenai County Commissioner,
lifelong Republican

"As a fan of Coeur d'Alene, history, and science, I found Mr. Irving's book delightful — the mix of all three of these topics is a great narrative that I hope everyone reads and enjoys. Our area — and our planet — is worth keeping."

Dan Gookin,
Coeur d'Alene City Councilman

LOSING HOME

GRIEF AND HOPE IN A CHANGING WORLD

Bill Irving

Copyright © by Bill Irving 2024

Interior and cover design by Jera Publishing
Photographs: All images in the author's collection
Edited by Suzanne Holland, MS

All rights reserved. No part of this publication may be reproduced, distributed, or transmitted in any form or by any means, including photocopying, recording, or other electronic or mechanical methods, without the prior written permission of the publisher, except in the case of brief quotations embodied in critical reviews and certain other noncommercial uses permitted by copyright law. For permission requests, write to the publisher, addressed "Attention: Permissions Coordinator," at the address below.

Bitterroot Mountain Publishing House, LLC
P.O. Box 3508, Hayden, ID 83835

For questions or information regarding permission for excerpts, please contact Bitterroot Mountain Publishing House at Editor@BMPHmedia.com

This is a work of nonfiction. Although the author and publisher have made every effort to ensure the accuracy and completeness of the information in this book, we assume no responsibility for error, inaccuracies, omissions, or any inconsistency herein. All slights of people, places, and organizations are unintentional.

Library of Congress Cataloging in Publication Data.

ISBNs:
978-1-960059-23-9 Soft Cover
978-1-960059-24-6 Hard Cover
978-1-960059-25-3 EBook

Printed in the United States of America

10 9 8 7 6 5 4 3 2 1

Contents

Acknowledgements   ix

Introduction   xi

PART I — **Home**

1. Casualties of Climate Change   2
   The story of Coeur d'Alene's first climate casualty, Fred Murphy, in the 80s, to more recent losses of home: clean air, unrestrained outdoor recreation, and a comfortable climate.

2. The Pull of Home   7
   The importance of home, how climate change is displacing people worldwide and the first-ever youth climate trial in Montana.

3. Gifts of Nature   23
   The qualities of home enjoyed by generations, provided by a stable climate. The pain of losing the home we've known while we still live here.

4. Ancestral Lands — 33
This area's original inhabitants, the Coeur d'Alene Tribe, their worldview, and their vibrant climate change program.

5. Coeur d'Alene's Climate History — 35
A review of the Lake City's major climate-related events, from 1910 to the present. Personal interviews with long-time residents who've lived through many of those events and the major changes they've witnessed.

6. Latest Climate Casualties — 64
From hotter summers since the 1990s to the harbinger of the future in 2015, plus the unprecedented heat wave of 2021.

7. Coeur d'Alene's Scary Climate Future — 75
Due to increasing greenhouse gas emissions, Coeur d'Alene's temperature is projected to sharply rise by 2100, would be "Lewiston North" by 2080.

PART II — Local Impacts

8. Local Climate Change Impacts — 80
From shorter ski seasons, diminishing lake quality, increasing wildfires and smoke, the physical and mental health impacts of smoke and hotter temperatures and the loss of social connections with the car culture.

PART III — Rising to Meet the Challenge

9. Common Sense Climate Change — 134
What is climate and how dumping 36 billion tons of heat-trapping gases into the atmosphere every year is causing bad things to happen.

10. United We Stand   143
   How an emergency can bring us together and the public's perception of climate change, including here in Kootenai County.

11. Who to Trust (and Not)   149
   From trusting the military about the dangers of climate change to the faith community, the investment world and the insurance industry. Who to trust and who not to.

PART IV — **Saving Home**

12. What You Can Do   162
   From cutting consumerism, driving less and walking and biking more, to harnessing renewable energy, eating less meat, planting more trees, investing wisely, developing climate adaptation plans and climate therapy.

13. What We Can Do Together   207
   Examples include establishing reliable cooling centers, protecting the community from wildfires, the power of movements, keeping fossil fuels in the ground, supporting historic state and national climate lawsuits, and describing important climate change-related federal legislation.

14. A Few Closing Words   249
   Willful ignorance, what we love about living here, living simply, joining others and the "incalculable return on investment" of educating girls.

References   261
Index   295

Illustrations

Figure 1 — Photo of grandfather Knudson

Figure 2 — Map of Kootenai County

Figure 3 — 1893-2020 Coeur d'Alene Annual Average Temperature

Figure 4 — Sled dog races on Sherman Ave., 1916

Figure 5 — Ice skating near Tubbs Hill, circa 1900

Figure 6 — Coeur d'Alene Eskimo ice hockey team

Figure 7 — Tractors clearing snow on Coeur d'Alene Lake, 1950

Figure 8 — Flag tied to automobile antenna, winter of 1968-1969

Figure 9 — 1970-2020 Coeur d'Alene Annual Average Temperature

Figure 10 — Northwest Summer Temperatures, Spokane, WA

Figure 11 — Kootenai Health admissions for asthma, June-August 2021

Figure 12 — Historic to 2100 Average Temperatures, Coeur d'Alene

Figure 13 — Photo of Merle Miley and friends cutting ice on Lower Twin Lake

Figure 14 — Historic to 2100 temperatures, nine Kootenai County lakes

Figure 15 — Spokane yearly snowfall, 1893-2023

Figure 16 — Western U.S. very large fires risk, 2041-2070

Figure 17 — Photo of 2021 Coeur d'Alene Lake wildfire smoke

Figure 18 — Kootenai Health admissions for asthma, 2015-June 2021

Figure 19 — Kootenai Health admissions for COPD, 2015-June 2021

Figure 20 — Marimn Health visits for asthma, COPD & heat illnesses, 2015-2023

Figure 21 — Professor Ruan holding photo of whitest paint

Acknowledgements

So many people contributed to making this book possible, more than I would have ever imagined. Early on, Craig Cooper provided the knowledge and wisdom to help better frame my thoughts. He was also generous with his time, reading an early version of the book, as did Eric Walsh. Craig also wrote the section on Common Sense Climate Change. Thank you as well to Mike Bullard, an accomplished and wise author, who was my mentor over six, monthly meetings. Mike came up with the title as well, which I immediately liked.

My two biggest supporters have been my identical twin brother, Richard, and my significant other, Ginny Taft. Richard read short early drafts, offering his honest opinion and encouragement. Along the way he developed an interest in climate change as well. Ginny was initially an interviewee, who developed into someone I love and appreciate. A published author herself, she taught me how to write shorter, more readable sentences, and was a wise and creative sounding board and editor-in-residence. She was eternally patient with my late night and early morning writing forays as well. Suzanne, my editor, helped with sentence structure and readability, as did Lisa Clark.

Thank you also to my father for his love of research and writing, and his interest in listening to climate change stories I read to him early on. Finally, I dedicate this book to my maternal grandfather. Grandpa's soft hand and warm heart, as well as his love of Coeur d'Alene, fostered my love for the Lake City and inspired the theme for this book. Grandpa was forever kind and patient with us rambunctious grandchildren. I loved him very much.

Introduction

What is home? It's more than a house. A home offers refuge, safety, comfort, as well as family.

What is home as social connection? It's our deep, emotional connection with family and friends, as well as our neighbors, co-workers and the greater community.

What is home as community? It's more than the people, where we shop or our kids go to school. It's the natural environment as well. It's our home's seasonal temperatures, when and how much it rains, snows, or how windy it gets. Over time we get to know what's "normal," within a certain range, for where we live. We come to rely on that predictability.

But that predictability, based on a stable climate, can be lost. Indeed, this is the story of my home losing its identity to climate change and what we can do about it. The loss of abundant snowfall, for example, on which we are dependent for our water, in lakes, rivers and streams.

For the past thirty-eight years I've lived in the greater Coeur d'Alene, Idaho area, where we're experiencing persistently hotter summers and choking wildfire smoke. Wildfires are more frequent, intense and longer lasting. Summer heat creates algae blooms in our local lakes. Lake and river levels are lower and warmer. Winters bring less snow and more rain than in the past.

As a result, we're losing our cool summer nights and clear summer skies. Once pristine swimming and fishing are marred by toxic algae blooms. Tourism, the lifeblood of our town's summer economy, suffers under the pall of heavy wildfire smoke. Many of us either flee or hide indoors from the dangerous smoke and rising temperatures. It's been years since we've had a truly snowy Christmas as well.

This is the first book focused on *local climate change,* where people live, day to day, written by a long-time resident, not a climate scientist. It highlights what people have witnessed over decades, and live with now, in their home community. Thoroughly researched and documented, these local impacts include interviews with more than thirty long-time residents, as well as young people.

Physical and mental health consequences of rising temperatures and wildfire smoke are described. The future of winter lake ice and nearby skiing in a rapidly warming world are discussed, among other topics.

Such local impacts need local solutions, tailored to actions residents can take, individually and collectively. Working with city and county governments, for example, citizens can improve the walkability of their neighborhoods — having daily amenities, such as grocery stores, restaurants, schools, and parks within a twenty-minute walk or bike ride.

Car use would be reduced, along with traffic congestion and tailpipe and greenhouse gas emissions. Such a long-term action, for example, would offer a return to the home many of us want, with opportunities to socialize with neighbors, go for relaxing, safe walks or bicycle rides. To have much of what we want and need nearby as well. Home would become healthier, happier and more connected.

INTRODUCTION

Important federal legislation as well as the latest on electric vehicles and heat pumps are discussed. Historic court cases, brought against state and national governments to secure a livable climate for children too young to vote, are reviewed as well.

Why my interest in this subject? I've loved Coeur d'Alene since I began swimming in the lake, during summer visits to mom's parents, while growing up. From sunup to sun down I swam with my brother and sister and visited Grandpa at his downtown feed store. Grandpa loved living in Coeur d'Alene. With its then cold, snowy winters and beautiful lake, I suspect it reminded him of his native Norway.

My motivation to write this book is to not only save the home of my grandfather, whom I loved dearly, and my own, but also to rekindle the passion in each of us to protect our home, no matter where we live.

Now to return to the greater Coeur d'Alene area.

An important aspect to those of us who live here, and its attraction to others, is the nearby lakes and a comfortable climate. Tourists flock here to enjoy the beauty of our lakes and the many recreational activities surrounding them. We love to hunt, fish, hike, swim and boat, as well as play golf and garden. Many enjoy the variety of our four seasons. Melting snow feeds our area's lakes, rivers and streams. The cost of living is relatively low as well.

How about our crazy weather, though? It's below zero one week and two weeks later it's fifty degrees, in late January! Wildfire smoke nearly every year is unlike our summers of the past, with beautiful blue skies. It's hotter in the summer, too. The heat wave in 2021, driven by climate change, was historic. The death toll exceeded 1,400, with at least 808 estimated in western Canada (Global Catastrophe Recap, 2021, p. 13).

For the first time that summer I was scared. Scared that at age sixty-nine I might perish in the heat, living in a mobile home without air conditioning. You may know someone who's had to go to the Emergency Room or been hospitalized for a heat-related illness in the past few years.

Wildfire smoke may have sent your next-door neighbor's child or grandchild to the hospital. They may have missed days of school, needed to use a rescue inhaler or had to see their doctor more often. Your grandparent or aunt or uncle's heart condition, or COPD, may have worsened due to the smoke.

A co-worker of mine, who has asthma, can barely breathe if she goes outside on smokey days. Imagine how panicked you'd be if you had to struggle to breathe. Chances are that someone you know has had to live with that fear. They may have become depressed as well.

Is it possible that all of this is the effect of global warming? Here in paradise? The following poem expresses our struggle with how the world is changing.

MY BASTION

I always thought I was safe
In this, my bastion,
Protected space.

There is nothing that I lack.
Strong mountain at my back.
Sunrise fills my eyes.
The light and mist
Kiss
And water flows in this
My verdant valley.

INTRODUCTION

Here I am impervious
To chaos.
Not that I don't care,
But the problems are just —
Elsewhere.
It is someone else's loss.
Or so I thought.

I heard a whisper.
Then louder
A rumbling
Of those bumbling
Malcontents.
They make no sense.
I heard them spew
Their points of view.
Could their predictions be true?

Or is it just a sick
And twisted trick?
I am so weary
Of conspiracy theory.
Those tree huggers again!
When will it end?
I then decide
They spread vicious lies.
"Hmph!" I said and rolled my eyes.

I thought I knew it all, you see.
"I am safe.
It won't touch me!"

(excerpted poem by Virginia Taft, 2022)

One of Coeur d'Alene's most notable residents experienced these changes firsthand, in 1986.

PART I

HOME

"To be rooted is the most important and least recognized need of the human soul."

Simone Weil,
French philosopher

CHAPTER 1

Casualties of Climate Change

The First Casualty

FRED MURPHY LOVED BEING on the water. At age nine he first helped operate his father's ninety-foot steamboat on the Pend Oreille River. His first job, at fourteen, was at the Coeur d'Alene Mill, piloting its tugboat. Felled trees were brought in off the lake to be cut for lumber. It was dangerous work — he lost two middle fingers doing it. Murphy, however, quit school in the eighth grade to do the work he loved.

"Agile as cats," he and his younger brother were paid $25 to walk on towlines between two steamers during Fourth of July shows at the Coeur d'Alene waterfront (Emerson, 1988, p. 8). During the area's worst flood in history, in December 1933, Murphy piloted a tug that helped to break up "a mass of floating material" (including whole buildings, railroad trestles and dead livestock) that threatened to take out the Blackwell bridge (Emerson, p. 33).

The following year he eloped, at age twenty-five, with eighteen-year-old Virginia Mason. They later moved to a large float house on the lake, and had the first of five children. As the children grew, Virginia ice skated across the lake after dinner to PTA meetings in Coeur d'Alene, two miles away (Emerson, pp. 42-43).

She recalled the beauty of one trip: "Unusual formations of hoar frost had covered the smooth ice, and a full moon had risen over the mountains...the crystals of frost had sparkled like diamonds in the glow of the moonlight" (Emerson, pp. 42-43).

Although celebrated as a tugboat captain, Murphy later developed a marine contracting business. A biography of his life, written by best friend Tom Emerson, noted his pile driving work in every bay in the early '50s "established his image as the Captain of the Lake" (Emerson, p. 115).

In 1972 Murphy and his son put in underpinnings for the Hagadone Headquarters building on the lake, in downtown Coeur d'Alene. Later, in April 1985 he completed building the Coeur d'Alene Resort's longest floating boardwalk in the world, as well as its multi-slip marina.

Sunday, January 12, 1986 should have been a celebration of Fred and Virginia's fifty-second wedding anniversary. Tragically, it was not. Fred decided to go for a pleasure ride on his snowmobile that bright sunny afternoon. The ice may have looked solid enough, but it wasn't. It was likely thawing and slushy underneath.

After a few hours, when she couldn't spot him on the lake, Virginia called their son, Skip. He rushed to the scene and spotted his dad's hat in the water. Skip quickly broke ice with the tugboat and found Murphy floating in the ice water. After he pulled him out, nearly falling in himself, Skip called the police. CPR and all other efforts were undertaken but he couldn't be revived. He was, his best friend related, "deep in the lake he so dearly loved" (Emerson, p. 1).

The world Murphy had grown up with was gone.

He had no idea about major changes in the world back then; none of us did. He was simply its first local casualty.

After his death, the deteriorated condition of the ice was highlighted. "Half of the ice is rotten, not just on Coeur d'Alene but all the lakes," said Kootenai County Sheriff's Office Sgt. Gary Anderson.

Coeur d'Alene assistant fire chief Mike Budvarson, said: "It is crappy everywhere. If the ice had been building up without snow on it, it would be fine. But the snow melts and the ice has what I call little worms in it. Even ice on Fernan Lake, for all its thickness, is rotten" (Tidball, 1986).

More Casualties, Years Later

During the summer of 2020 my phone began to display something I'd never seen before: the letters AQI, followed by a number. I had no idea what that meant, so I ignored it. When it continued, though, I found out that it stands for Air Quality Index, a measure of how healthy or unhealthy the local air is. That air quality is reported as a number, from 0 to 500. The higher the number, the worse the air quality, with increasingly serious health consequences.

Why was our community's AQI score on my phone? Wildfire smoke had made our local air unhealthy to breathe, especially for those sensitive to air pollution. Tabitha Day, my co-worker at the time, is a prime example. Plagued with asthma, she couldn't go outside in the smoke without struggling to breathe. The air had gotten so bad, she told me, that she and her husband were considering moving away.

Although I was in good health, I stayed indoors as well. My eyes still stung, though. For nearly two weeks I couldn't see the nearby hills through the gray haze. My mood darkened. Drearily, I asked myself: "Is this ever going to end?"

After parts of three of the past four summers were erased by wildfire smoke, I realized a second quality of my home had been stolen: ***clean air.***

Our unrestrained enjoyment of outdoor recreation has been a casualty as well.

The joy of swimming all day in our local lakes, as I'd done as a youngster, is being robbed by toxic algae blooms and warmer waters. Property values drop near those foul-smelling blooms. Fishing suffers as well. Fish suffocate in such oxygen-deprived waters, and native, cold-water fish can't survive in too warm waters (replaced by invasive species). These and other forms of outdoor recreation, part of our envied outdoor lifestyle style and ingrained in our cultural identity, are being lost.

Such blooms are dangerous, actually a cyanobacteria that sickens and kills fish, as well as dogs at high concentrations. They're formed in warm, shallow, oxygen-deprived waters along with phosphorus and nitrogen from fertilizers.

Here in North Idaho, we've experienced more such blooms. The summer of 2021 was historic, as eleven water bodies were issued health advisories, the most ever (Higbee, 2021 email). These yearly health warnings, and nearly annual wildfires and smoke, restrict our enjoyment of outdoor recreation. No such restrictions existed here fifty years ago.

Local lakes, rivers and streams are heating up, increasing evaporation as well. The result: lake levels and stream flows are dropping, altering fish populations and aquatic ecology.

We're also losing our ***comfortable climate***, particularly during our renowned summers. Hotter temperatures were documented by *Coeur d'Alene Press* meteorologist, Randy Mann, in a March 2020 *Press* column.

During the decade of the 2010s, Mann found combined high and low summer temperatures were 68.76°F. Two decades before, in the 1990s, the June through August temperatures were nearly three degrees cooler — 66.08°F (Mann, 2020). Our prized, cooler overnights are slipping away as well. Sleeping is becoming more difficult and dependent upon air conditioning to stay cool.

These and other casualties are causing us to lose what we love about living here. It's so gradual we hardly notice it until we reflect back on how conditions were before.

If it's climate change (or global warming) that is causing these changes, what is that? Think of the atmosphere as a blanket surrounding the earth. When we burn coal, oil or natural gas for energy, we thicken that pollution blanket. The thicker the blanket, the more heat is trapped underneath. As a result, the earth is overheating, which we feel here, and throughout the world.

CHAPTER 2

The Pull of Home

> "No matter who you are, or where you are,
> instinct tells you to go home."
> Laura Marney,
> Novelist

> "People exploit what has a price or what they conclude to be merely of value; they defend what they love. Love cannot be priced... we love what we particularly know."
> William J. Lines,
> Australian essayist
> ("Money", p. 26)

I ASKED MY AUNT AND uncle how my grandparents had felt about living in Coeur d'Alene. Aunt Myrna replied without hesitation: "They wouldn't have lived anywhere else." Many of us feel the same way about where we live.

It likely will come as no surprise that the majority of people worldwide feel attached to their places of residence. In a worldwide survey of twenty-four countries, participants were asked to rank places to which they would be willing to move if it meant an improvement in their life condition. The respondents' own neighborhood ranked first in all countries, followed by another neighborhood in the same city (Laczko, 2005).

A home is more than four walls and a roof. A house is made into a home by actions of its occupants to make it their own. A home is also a feeling, captured in the expression "Home is where the heart is." It's our emotional connection to the people we love, where friends and family live or gather. We raise our children from home. Home is where they return for the holidays.

Home is also part of our identity. We're known by how we take care of inside and outside our residence. We introduce ourselves as being "from" this town or state. Soldiers fight to defend their *home*land.

When asked "What does home mean to you?" one subscriber to the magazine *Real Simple* noted that it's more than a place: "Home isn't a place, it's a feeling. A cozy, warm bed, being in the loving arms of a husband, the sound of your husband's key in the door at the end of a long day, or the smells of home. Home is sanctuary." "A sensation of peace on a cozy, rainy Sunday," another wrote. Or it is "the relief when you pull into the driveway after a long trip."

Another subscriber summarized: "Houses get bought and sold; a home stays with you always."

It has with me. Since I was three months old, actually, when Mom first brought my identical twin brother, Dick, and I to see her parents in Coeur d'Alene. For the next seventeen summers, along with younger sister Janet, we swam in the lake every day, two blocks from their house. My love of Coeur d'Alene originated in those summer visits, filled with play, adventure, security and love.

Erupting from the car after a long drive from Sacramento, the three of us kids scrambled excitedly into Grandma and Grandpa's house. After giving and receiving obligatory hugs, the three of us would scurry upstairs to change into our swimsuits. Towels draped over our shoulders, we raced to the beach, Mom in tow.

Dick and I immediately plunged into the water, thrilled to be swimming in "our lake" again. We dove for anything we could find on the bottom, not far from the shore—cans, bottles, occasional shoes and the like. Janet most often swam near shore and built sand castles. Mom visited with folks and read. Dad seldom joined us, working on projects with Grandpa at his store.

When it was time for dinner, we'd scurry, like Mexican jumping beans, on the hot sidewalk, back to Grandma and Grandpa's house, the cool, green backyard lawn a welcome relief. With sand between our toes, we toweled off on the back porch and steps.

Dick and I often tried to sneak cookies from Grandma's cookie jar before dinner. Slowly I'd pick up the lid, grab two or three oatmeal cookies and try to set the lid down, oh so quietly. But Grandma always heard me: "Is that you, Bill?" she'd ask from the living room. "Darn! Caught again" I'd mutter under my breath, then answered yes, sheepishly. Grandma half-smiled as I walked into the living room and headed upstairs to change out of my swimming suit.

When we weren't swimming, we enjoyed visiting Grandpa at his feed store on 3rd Street (now a large parking garage!). Pushing each other around on hand trucks, more than once we accidentally ran into fifty-pound bags of rabbit pellets, splitting them open. Ever so kindly, Grandpa would say, "Whoa, be careful," then patiently tape the hole closed, and we'd return to playing. Only once did he raise his voice, to protect Dick from getting hit by a car, he remembers. Mom reported the same growing up.

I loved Grandpa. He remains the kindest, most patient man I've ever known. He regularly took us to The Missouri Kitchen (known as Hudson's Hamburgers since 1963) for our staple of pickle and onion hamburgers and sodas. Grandpa drank buttermilk, a strange concoction to us. When we finished, the counter person would ask: "How much, Hal?" Grandpa would add it up in his head, then pay and we'd leave. His word was trusted.

During the Great Depression, he extended credit to farmers and ranchers, dependent on his feed for their livestock and seeds to grow their crops. Although he was generous, he was no pushover. When a farmer attempted to renege on what he owed, Grandpa, a former football player, tackled him outside the store. That's a favorite family story. While it may seem out of character for such a kind man, it wasn't. Integrity was important to him.

Grandpa was on the first Board of Directors of what would become the member-owned Kootenai Electric Cooperative. Its first office, I recently discovered, was in Grandpa's store (KEC Annual Report, 2012, p. 2, p. 4). Established following passage of the Rural Electrification Act in 1936, KEC today provides electricity to our cabin on Lower Twin Lake and other locales.

With a medium build and a shock of white hair (Figure 1), Grandpa smelled of Old Spice aftershave in the morning. He wore thick, perpetually smeared glasses. His daily uniform consisted of black or dark brown pants, with suspenders, inches above the ankle. Heavy black boots and a checkered red and gray long-sleeved flannel shirt completed the ensemble. Lava soap was a nightly companion. He regularly ate oatmeal for breakfast.

Once each summer visit, we returned sold flats of flowers to Spokane nurseries and purchased more store supplies in town. Janet sat in the cab of the International truck with Grandpa, Dick and I in the back between two, eight-foot-high wooden racks. We wore jackets, shivering in the brisk early morning air. After the nursery stops, we picked up our four cousins in the Spokane valley. Grandpa brought jelly-filled doughnuts.

The seven of us were Grandpa's "helpers" at the downtown Boyd-Conlee feed store, and other stops. We each received a nickel or two for the soda pop and candy machines. Chasing each other through the aisles, "helping" Grandpa was a fun adventure.

We always ate dinner at the house. Grandma was a good cook, assisted by Mom. The food was healthy, delicious and there was plenty of it: corn on the cob, chicken, spaghetti, mashed potatoes and salads. Occasionally we had cookies or ice cream for dessert.

After dinner, we joined Mom in an assembly line to wash dishes. Mom washed, then passed to one of us who'd rinse, another to dry, and the last would put the clean dishes away.

For a treat, we headed to A & W for root beer, north on Government Way, at what was then the far north end of town. Kids sat in the back, on the Chevy New Yorker's bouncy seats with Dad. Mom and Grandma were up front. Grandpa drove. Impatient, it always seemed to take forever to get there.

Each of us remained in the car as Grandpa ordered from a girl attendant on roller skates. Big glass mugs of foam-topped root beer arrived on a tray. Two hooks straddled the rolled down driver's window. Grandpa passed around the mugs. The root beer was deliciously cold and tasty on a warm summer evening. We devoured ours in no time.

Having been raised on a ranch, Grandpa checked the barometer on the living room wall every evening before dinner. I'm convinced his keen appreciation for the weather lives on in my passion to protect our home from the ravages of climate change.

Grandpa loved living in Coeur d'Alene. It's beautiful lake, deep snows and cold winters likely reminded him of his native Norway.

Often, into his 70s, in good weather, he walked to and from his Fort Grounds house to work, a mile each way. He enjoyed its small town friendly and safe feel.

It's painful to witness global warming dismantling the home he loved so dearly, as well as my own home since 1986. The climate had been stable where we grew up, first in Sacramento until 1965, then Ellensburg, Washington until 1974 and during the summers in the Lake City until 1969.

"Home is a shelter from storms — all storms."
William J. Bennett

Figure 1

THE PULL OF HOME

None of us saw the end coming for our grandparents. At age eighty-five, Grandma had surgery for a bowel obstruction. A week later she developed sepsis and died, in June 1969. She and Grandpa had been married for thirty-eight years. I had no idea at the time what her loss meant to Grandpa.

Following Grandma's funeral (which we didn't attend, likely due to being in school), Mom asked to look at Grandpa's finances. We all gathered in the living room to hear the news. Mom was blunt. He couldn't afford to keep the house or likely his feed store, she said. Immediately, tears welled up in Grandpa's eyes, something I had never seen before. He pleaded with her: "But it's all I have!" I felt scared, then desperately sad and helpless to assuage his pain.

We stayed with Grandpa for the remainder of the summer. He was not the same though, drained of energy, sad, grieving his loss and uncertain future. A maid was hired to clean and cook when we had to return to Ellensburg for school and work. We visited nearly every weekend.

The loss of his beloved wife and the prospect of losing his house and likely his store, however, was more than he could bear. One afternoon in October he committed suicide, spreading rat poison in a confined space in an upstairs room next to his store. When he failed to return home from work, the maid called Uncle Bob, in Spokane, who found him dead in the loft.

The funeral home was packed for the service. I was somber, then annoyed that the family had to sit behind a sheer black net. I wanted to see the faces of the people who'd come to pay their respects to Grandpa but I couldn't. The children were told to be quiet during the funeral, Janet remembers.

The kindest, most patient man I'd ever known was dead at age eighty. The joys of summer for us three grandchildren died as well.

When he passed, the Lake City was a sleepy little town of 17,000, blessed with natural gifts. It would be seventeen years until Fred Murphy perished.

Climate change was yet to strike.

Accelerating climate change

Fifty years after the death of my grandparents, the situation is alarmingly different. Wildfires rarely seen before threaten, most recently the 3,000-acre Ridge Creek fire northeast of Hayden Lake, burning through the month of August in 2023. Spokane County fires as well took weeks to contain, then extinguish, with losses of life and homes.

While this is no longer my childhood home away from home, climate change has become a concern for young people, both here and across the country. A study published in the medical journal *The Lancet* in 2021 found nearly half of Americans surveyed from ages sixteen to twenty-five were very worried or extremely worried about climate change and thought it doomed humanity (Hickman, et al., 2021).

Sage Pedersen and April O'Quinn are local 17-year-olds, the same age as when I swam without a care in Coeur d'Alene Lake. I met Sage at a local restaurant in mid-August. She described being "scared" one summer day when she jumped into Hayden Lake and the temperature was the same as the air temperature. No longer did the lake provide the relief from heat she'd come to rely on.

April emailed me during a layover in her family's hiking trip near Lake Tahoe to say her interest in the climate began in elementary school. April sold homemade cookies outside her house, then donated one hundred percent of the proceeds to a nonprofit dedicated to researching melting ice caps in the Arctic. Nowadays, she states: "I am pained to see local hiking trails torched by fire."

THE PULL OF HOME

In the summer of 2021, April marked the water level on the shores of Hayden Lake, worried about how low the July water level was. An avid skier at Schweitzer Mountain as well, she worries if skiing "will become a hobby of the past, lost to climate change."

That same reticence extends to her having children: "Personally, I wouldn't be comfortable bringing a child into a world where their future survival is undetermined. It simply doesn't seem fair to leave a new generation with endless climate issues that will haunt them," she volunteered.

April referred me to Owen Finn, also a 17-year-old senior to be at Coeur d'Alene High School. We met on an early Friday morning in late August at the Bakery By The Lake restaurant. A native of the Lake City, Owen shared his interest in majoring in climatology or geography at Oregon State University next year.

I asked Owen, "What changes in the weather have you noticed here?"

"The smoky season, in August, has become unbearable," he replied. The wildfire smoke began when he first attended the Kootenai County Fair at age 10, and has "slowly progressed" until now it's "super smoky and super hazy, not fun to be outside in." His local rock climbing is interrupted in mid-July by too hot and too smokey conditions as well.

Climate change "will definitely be a factor" in whether or not he has children: "I'll see how this plays out," he said, depending on whether "people take the initiative to stop it or not."

"Overall, how much of a concern is the climate issue to you?" I asked.

"Moderate." he said. "An attempt to stay moderate, to not be too fearful: If you're too fearful, you can't do anything."

Next, I share how our connection to home is even deeper than most realize.

Displacement

"No one leaves home unless home is the mouth of a shark."
Somali-British poet

The above quotation refers to fleeing home, as hundreds of thousands of refugees have had to do to avoid persecution, violence, starvation or war in the past six or seven years. Many are fleeing hot places as well. The U.N. Refugee Agency, UNHCR, says "... 70 percent of refugees and 80 percent of internally displaced people originate from countries on the front lines of climate change" (Refugees, U.N., 2023). We call them the displaced.

However, some anthropologists are now saying that displacement is a global condition, affecting people wherever they live, not just those on the move. Australian anthropologist Hedda Askland and colleagues state that displacement needs to be focused on the relationship between person and place, on the experience of "feeling at home."

Dis-place-ment, she writes (Ramsay & Askland, 2020), expresses "the loss of a place called 'home,' a place in which lives are fulfilled with existential meaning" (Askland, H.H., et al., 2022). More on existential meaning shortly.

Consider the example of bushfires in Australia during the summer of 2019-2020. Askland points out at least thirty-three people were killed and 400 others died due to the poor air quality. Over one billion animals were killed. Those who survived in the beach hamlets and rural townships had their lives and homes "ruptured." Their sense of security with the past and future was torn asunder, in their own homes. They are victims of climate change induced displacement.

Displacement "involves having one's sense of purposeful being and purposive connection to place, time, and social worlds ruptured"

(Askland, H.H., *et al.*, 2022). Nothing is as it was, nor a future life as it was imagined is possible any longer. Loss and grief are everywhere. One's way of life, and authority over it, as well as one's future, are lost, Askland states. Social relationships are forever transformed.

Displacement has become a worldwide condition, Askland asserts, "imposed by external forces, outside of individual control", such as by climate change. *Increasingly, the mouth of a shark is everywhere.*

Fortunately, most of us have not had to face a climate change induced disaster here. However, we've had to endure searing impacts from wildfires, choking smoke, heat waves, ongoing drought, diminishing snowfall and its early melt, and the loss of winter lake ice.

A destabilized climate makes life more precarious for all.

Crisis Next Door

In June 2023, Montana was the site of the first ever constitutional climate trial. Sixteen young people, ages five to twenty-two, sued the state of Montana for violating their right to a "clean and healthful environment," enshrined in the state's Constitution. In *Held v. State of Montana*, along with expert witnesses, they testified to the many harms climate change had caused them and their state.

For five days, the sixteen children shared their worry of losing "beautiful places" such as Glacier National Park, or their fears when rivers ran dry or flooded their homes. Also, of becoming anxious or depressed when they were forced to isolate indoors due to wildfire smoke and excessive heat. Two young women were reluctant to bring children into the world in these conditions.

At the end of the week, in a park near the courthouse, nineteen-year-old Grace spoke for the youth. The weight of young people having to worry about their futures in an increasingly climate disrupted world hung in the early evening air. Their courage was applauded by the audience.

After thanking the gathered crowd for their support, holding back tears as she quickly left the microphone, a somber Grace said: "And I wish we didn't have to do this."

The circumstances described by the young people in that Montana courtroom are similar to what many of us have experienced in the Coeur d'Alene area.

Rikki, the lead plaintiff, now twenty-two, was eighteen when the case was filed in March 2020. Her family has a hunting business on a 3,000-acre ranch in southeast Montana. Road closures, due to wildfires in August 2021, during peak season, cost the family half their revenue. The financial toll on her family registered in Rikki's stifled tears. She suffered headaches and heat exhaustion from working on their ranch during the wildfires as well, she said.

On the stand, Grace shared the following feelings about the effects of climate disruption in Montana: "Frustration. Guilt. Loss. Anxiety. Grief…. The fear of loss of beautiful places." In 2014, her father had to pull the raft on a family float trip on the too low Clark Fork River. About having children, Grace echoed another young plaintiff: "I'm not sure, medically or ethically, that I want to have children."

Now seventeen, Eva spent seven hours filling sandbags in 2022 during a flood near her home in Livingston. She was "very, very scared" that the levee might fail. In 2018 a spring flood took out the town's main bridge. A temporary bridge was then washed away, forcing a much longer commute to town for her and her family. Eva said she fears "losing control and a sense of hopelessness" over the climate situation in the Big Sky state.

Ruby, fifteen, and Lillian, twelve, are Crow Tribal members. Their father, Shane, spoke for them during the trial. The annual

Crow Fair is an important part of the Tribe's and the girls' spiritual and cultural identity, he said. Dancing, a rodeo and powwows are held during the third week of August. Recently, due to over 100°F temperatures, torrential downpours and too smoky conditions, the rodeo had to be postponed — a "huge disappointment" for the girls.

In middle school, Claire helped raise $125,000 to put solar panels on her school. But since Montana restricts solar to only fifty kilowatts per facility, the money provided power to only one-quarter of the school. Now, as a winter ski instructor at Big Sky Resort, later snowpacks have reduced the number of skiers, she said, and thus her income. Born with an undisclosed disability, Claire noted her "only therapy is outside." However, climate driven wildfire smoke and excessive heat have disrupted her ability to be outdoors.

Finally, Olivia, now twenty. Diagnosed with exercise-induced asthma in the seventh grade, triggered by smoke, she said she could barely breathe during wildfire smoke in May 2023. In addition, her eyes swell shut due to "extreme allergies," exacerbated by increasing heat and smoke. Through tears, she described her climate-induced anxiety as a "giant weight on my chest...I would not want to bring a child into that," she added.

The trial highlighted Montana's hold on its fossil-fuel based energy system. In 2011, the state prohibited state agencies from considering greenhouse gas emissions in the permitting of energy facilities. In addition, over thirty years, its Department of Environmental Quality had never denied a permit for a coal, oil, or gas operation (Findings of Fact & Conclusions of Law, July 2023, p. 15).

In mid-August, the court ruled that the state's oil and gas policies were unconstitutional. The ruling was considered historic. Judge Kathy Seeley found the state's prohibiting the consideration of greenhouse gas emissions and climate impacts in the energy permitting process to be "facially unconstitutional" and infringed on the young people's constitutional rights to a safe environment (Bloomberg Law, August 14, 2023).

"I think this is the strongest decision on climate change ever issued by a court," said Michael Gerrard, director of Columbia Law School's Sabin Center for Climate Change Law. "Putting a human face on this global problem worked well in this courtroom, and may well be followed elsewhere," Gerrard said in an email to the article's author (Bloomberg Law, August 14, 2023).

Lawsuits that use a safe climate as an affirmative human or constitutional right, as in this case, are gaining credence in this country, the article noted. For example, in March, the Hawaiian Supreme Court recognized a human right to a stable climate in a ruling against a biomass power plant developer (Bloomberg Law, 2023, August 14).

The win in the first-ever climate trial in Montana is historic and precedent-setting. Along with a federal climate lawsuit, described later, the noose is tightening on the out-of-control fossil fuel industry and governments worldwide. But will the noose close quickly enough to save each of our homes? We must have an affirmative answer before 2030.

The fossil fuel industry will fight each step of the way, though. In a case involving a natural gas power plant outside Laurel, Montana, the state requested to pause Judge Seely's ruling. In January 2024 the Supreme Court denied its request (Miller, 2024). In February, though, a district court judge ruled that construction of the plant could continue (Lutey, 2024). A date for oral arguments in the Supreme Court appeal of the *Held v. Montana* decision awaits, on July 10, 2024.

The climate burden we have foisted on young people has been calculated. "To compensate, the lifetime carbon dioxide emissions (or carbon budget) of the average young person today will need to be *eight times less than that of their grandparents* to restrict global warming to 1.5°C [2.7°F], the limit set out by the Intergovernmental Panel on Climate Change in 2018" (Hausfather, Z., 2019).

Introduction to Coeur d'Alene

Coeur d'Alene (pronounced kore-duh-LANE), is a combination of three French words. Coeur is the French word for heart, Alene means awl, a pointed tool to pierce leather. The d' means of.... Thus "heart of awl." It's generally agreed that the name describes the native Coeur d'Alenes sharp trading practices ("sharp" as an awl).

> French-speaking Iroquois Indians were first retained by David Thompson of the North West Fur Trading Company as guides and scouts. In addition, Lewis and Clark described a meeting between traders and the Coeur d'Alenes at a Nez Perce camp during the 1805 expedition (Dahlgren & Carbonneau-Kincaid, 1996, 2).

The Lake City itself is located two and a half hours south of the U.S.-Canadian border, and thirty miles east of Spokane, Washington. It's the county seat of Kootenai County, shown in Figure 2. Many lakes, rivers and streams dot the county. Approximately 56,000 people live in Coeur d'Alene today. In April 2021 the town made the headlines as the #1 emerging real estate market in the country (Edelen, 2021).

Coeur d'Alene has a temperate climate, with four seasons. The average snowfall is 69.8 inches per year, but declining. Summers are hot, but historically not brutal (although hotter lately). While winter low temperatures are regularly below freezing, lows below 20°F don't last long. Enough rain in the spring helps to green the town's trees, grasses and annuals. The Lake City's elevation is 2,188 feet.

Floods, from heavy rain and melting snow, are "the most common natural disaster in North Idaho." They've been destructive in the past—the last major flood was in 1997 (Kootenai County floodplain letter, 2023).

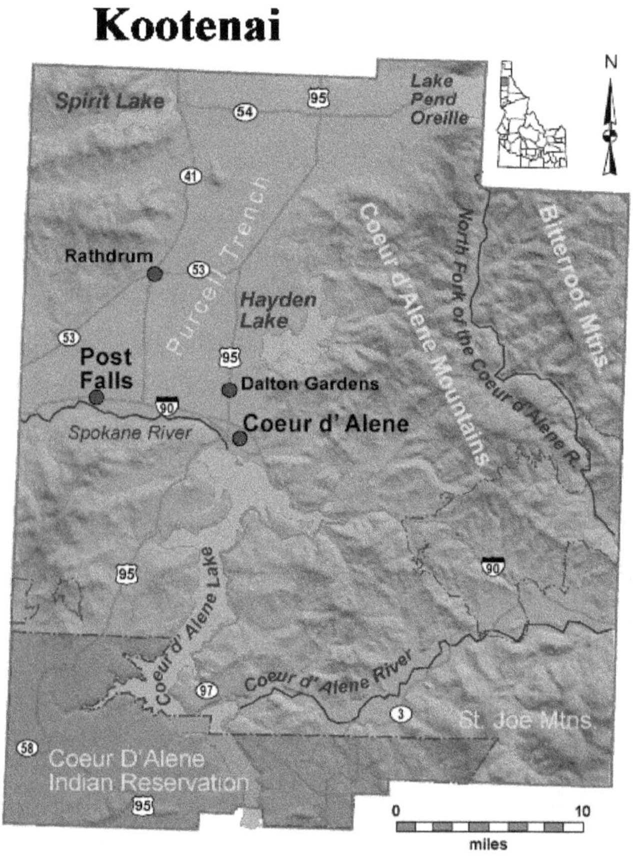

The Digital Atlas of Idaho

Figure 2

CHAPTER 3

Gifts of Nature

"Climatic conditions in Kootenai county are as favorable to health and to the production of crops as anywhere in the Northwest."
An Illustrated History of North Idaho, p. 820

FOR MORE THAN 100 years, the Coeur d'Alene area has been known for its beauty, comfortable climate and outdoor-focused activities. These and other qualities of home I refer to as its "gifts of nature." They were first extolled in writing in 1903 in *An Illustrated History of North Idaho*, when Kootenai County consisted of five counties, not one.

- Kootenai County's "pleasant winters."
- Its bountiful snow in the nearby mountains — "There is always an abundance of snow."
- Its comfortable summer temperature — "summer is never oppressive"; "almost every town in the county is a delightful summer resort."
- An outdoors/sportsman's paradise — "A paradise for hunters, fishermen and tourists."

- Its beauty — "The valley and mountain streams and lakes possess a rare beauty." (*An Illustrated History of North Idaho*, p. 820).

These attributes are our area's calling card. Like the gold rush, once discovered, throngs of visitors are attracted by its gifts. Some return as residents, others who left are called back by these blessings.

To find out how much our home's gifts of nature are present for people today, I asked members of a local writing group what they appreciate about living in the Coeur d'Alene area.

Only one member was a lifelong resident of the Lake City. In her early thirties, she noted loving "the mountains and forests and the lake that surround[s] the city." She appreciated as well "the dry climate with summers that don't get humid," plus "fairly mild" weather and the ability to experience "all four seasons." In addition, "spring crocuses and autumn leaves as well as sunny summers and white Christmases."

Another writer, formerly from Tucson, Arizona, raved about "coming home" to "amazing" crocus or hyacinths that were "visually stunning." She and her husband "love the trees too. It is soooo green compared to Tucson." The lakes and state parks drew high praise as well.

Four seasons were appreciated by another, specifically when "winter isn't too cold or too long." Living here nearly four years, in her mid-thirties, she added that "Coeur d'Alene Lake is playful, hiking is plentiful and the people are friendly."

A writer who's lived here for thirty years shared that "I am not much of a nature/outdoor person, but I do still like how many lakes there are to drive around." She noted it was important as well that Coeur d'Alene has grown, to provide opportunities such as the writing group.

GIFTS OF NATURE

The four seasons, especially winter, are big for another writer who's lived in North Idaho most of her life. She shares that she "exclaims over the snow, wishing it would come and go swiftly — I embrace its existence. The beautiful blanket of peace that descends upon field and stream and forest in the dead of winter time."

Another resident for the past sixteen years loves its peace and quiet: "Growing up in a metro area, I love that I can quiet my mind here. The noise, even though it's increased, is still quieter than other places. In general, it's a slower life than where I began."

A male resident moved here from Seattle in 2020. He likes the "smaller town feel and a better area to raise a family.... Less crime.... The lake and the mountains are beautiful. I like that we get regular snow in the winter and more days of sunshine and warmer temperatures in the summer."

Finally, a female writer who's lived in Coeur d'Alene more than forty years also appreciates its four seasons. In particular, she wrote, summer's "warm days and cool nights and lots of outdoor activities, including hiking, biking and kayaking," as well as its "natural beauty" and "friendliness of its people." In addition, she welcomes "Few violent storms, wind or poisonous things."

While a small survey, these comments likely represent the gifts of nature many people appreciate, even love, about living here: its beauty, lakes, rivers and streams, plus its four seasons, including its winter snowfall. In addition, it's provided a comfortable climate and a variety of outdoor activities

Some would say, with good reason, that Coeur d'Alene Lake is its own gift of nature. After all, it's listed as one of the most beautiful lakes in America (Vacasa, 2023). It's no wonder that Coeur d'Alene is also known as the Lake City.

Nor is it a surprise that the town's population has more than tripled, from 17,000 in 1970 to some 56,000 today. Once businesses began to capitalize on this area's gifts of nature, the Lake City grew rapidly. Its population increased twenty percent, from 2010 to 2020.

The bloom, however, has begun to fade. Our home is no longer the one long-time residents have known. Consider a few recent comments by local citizens. We'll begin with my first wife, Shelley, whom I interviewed in late August 2022, at a horse stable east of Fernan Lake.

Shelley has a passion for riding horses since age two, she said. At age four she put a front door mat over a split rail fence, tied a rope to the fence post and got on to ride. "Riding the rails" she laughingly called it. Since 2010 she's had her own horse. A few years ago, she began providing riding lessons, primarily to young girls, "to support their dream of riding."

Riding the rails isn't as easy anymore, though. "I need to squeeze people into the cooler part of the day [for riding lessons]," she said, "which eliminates the late afternoons and even the evenings now. It's just too hot."

The summer of 2021's climate change-driven heat wave also more than doubled the price of hay for the horses, from $225/ton to $500/ton. Even getting it was a big challenge, she reported. Shelley also lamented the "almost necessity" of air conditioning now; opening the windows at night doesn't always cool down the house.

A foot to a foot and a half of snow on her front yard used to last the entire winter in 1986. Now, well before the winter is over, she can see the grass. She and second husband, Ron, also use the snowblower "less and less in the winter."

In late August of 2022, I spoke to Paul Hanna as well, in the Hayden library. The uncle of a young man I'd interviewed in June, he shared his passion for flower gardening in Athol, twenty miles north of town. The heat has required that he use more water to sprinkle his garden, he said, adding: "We've had twenty-something days over 90° already, and we're not done yet." He couldn't remember this many days over 90° in the past as well.

When Hanna's family moved to the small town of Clark Fork (east of Lake Pend Oreille, north of Coeur d'Alene) when he was fourteen, in November 1960, it was three or four degrees below zero. Lake Pend Oreille at that time would "consistently" freeze, resulting in "great ice fishing."

Not anymore, though. Coeur d'Alene Lake doesn't freeze enough for people to drive on nowadays either. It's been "a long time" since it was safe enough to even walk on, he recalled.

In the waning days of December 2022, in between pickleball games, I interviewed Nils Rosdahl at the Kroc Center. He and wife, Mary, moved here from Seattle in 1980, the day Mt. St. Helens erupted. Rosdahl said he hasn't seen many changes in the climate, except needing to use more water to irrigate their garden. And no, he isn't concerned about his grandkids growing up in a global warming world: "I probably should be, but so far, it hasn't affected our lives."

I met my next interviewee at the previously mentioned writer's group meeting. "I've seen a lot of changes" regarding global warming, Virginia Taft said, since she arrived here in 1978.

The 2021 summer heat made it "too hot" for outdoor recreation. Until then she hadn't felt the need for air conditioning in her home, but that summer "it got hot and didn't cool off." The heat also "isolated the elderly in their homes." The disrupted climate has also worsened the local summer fire danger — "restricting our freedom" and making us feel "more fear-based."

Taft noted two unique impacts of local global warming as well: "Tornadoes have been reported and several episodes of "pretty violent winds," which tore down her neighbor's pine trees, caused a power outage and extensive property damage.

I met with Roger Brockhoff, an avid, longtime fisherman, at his Lake City home in November 2022. A native of Coeur d'Alene, he says the number of winter fish caught in Fernan Lake has plummeted, from twenty to forty crappies per day to "very few" nowadays. The previous three winters the ice hadn't been thick enough to even attempt to ice fish there. He hadn't noticed any difference in the taste of the fish due to climate change-induced warmer water temperatures, though.

Brockhoff doesn't fish if "it's terrible smoke," he said, but climate change is just "part of our life," like field burning on the Rathdrum Prairie, which ended in 2006.

These and other later interviewees have lived in this area for many years, some their entire lives. They've come to know its long-term weather, its normal range of variation in temperature, snow, wind and rain.

Unconsciously, over time, we all come to rely on those long-term patterns. We know, for example, about when it'll be warm enough to plant flowers outside and swim in the lake. Also, how hot it usually gets in July and August, when the first frost will hit and about how much snow we'll get, starting about when.

Without realizing it, we all depend on our climate being stable and predictable, within a certain range. Why? Because those conditions give us a sense of control in life, that life is not overwhelming, out of control, or frighteningly random. Climate stability has allowed us to lead more emotionally stable, predictable lives. To plan ahead as well.

When, however, the climate is disrupted, when what we've come to know and rely on abruptly changes, we become anxious. We feel vulnerable, alone, even helpless in the face of an out-of-control climate. It's scary. What was once solid ground is now shifting. We worry: if our home's climate is no longer stable, how can we organize our lives (e.g., plant crops and flowers, go on vacation, etc.)?

The impact, though, is even deeper. Without a dependable climate, the home we've known here—what makes Coeur d'Alene unique—would be lost, its essential nature extinguished. Our town's vibrant character would vanish, becoming nothing more than a name on a map. A house, no longer a home.

There is a nearby example of this process—Glacier National Park, in northwestern Montana. Due to climate change, the park is losing its namesake glaciers. Summer melting is outpacing winter snowfall, rapidly melting its glaciers. Of the 146 glaciers that existed in 1850, according to Dr. Daniel Fagre of the U.S. Geological Survey, only twenty-six survive today (Expert witness testimony of Dr. Fagre, 2023). They could all be gone by 2030 (Earth Observatory, NASA).

Without its glaciers, the heart of Glacier National Park will cease to exist. What took Nature millions of years to create we will have extinguished in less than 200 years. In different, but unmistakable ways, global warming is doing the same around the world.

There is a personal analog as well. Each of us has an essential nature, what uniquely animates us. My partner, Ginny, for example, has an abiding acceptance of others and life. She makes room for how people are, engaging them with ease. Shelley, my first wife, does as well. Both women's social ease and acceptance of others are part of their essential natures.

If either were to be afflicted by Alzheimer's, however, their essential natures would be lost. There would, tragically, be "no one *home."* Their identity, who they are to themselves, to others and the world, would be gone. In much the same way, home, as the source of our existential meaning in life, would be lost. There's a name for this home loss. A strange one: solastalgia.

Solastalgia

Solastalgia describes a disruption in a person's sense of home *while they still live there.* It's a homesickness, a sense of grief when one has lost what they love about where they live. A unique kind of trauma, it combines *solas,* as in *solace* (comfort in the face of distressing forces) and *algia,* which means pain, suffering, or sickness.

The inventor of the concept, philosopher Glenn Albrecht at Australia's University of Newcastle, summarized it this way: "In short, solastalgia is a form of homesickness one gets when one is *still at home"* (Albrecht, G., 2012). [author's emphasis] It involves a sense of dispossession and loss—the home you're living in is no longer the home you've known.

Often considered the "psychic pain of climate change," it can occur from both natural processes such as drought and forest fires or human-induced processes such as climate change and urbanization. It can lead to depression as well.

Solastalgia speaks to the power of place in the human psyche and how any significant disruption in where we live can be deeply troubling. Just ask any of those forced to flee their home from hurricanes, infernos, wildfire smoke, or floods. Being torn from one's home, and the deep relationships built there, can leave lasting psychological scars.

Local examples are more subtle: being unable to ice fish in that favorite spot due to precariously thin ice. Being upset that the snow you enjoy watching quietly float to the ground is too often replaced by pelting rain during the height of winter. Seeing puddled, frozen rain water on the ground, rather than inches of beautiful white snow, disturbs the home we've known.

Other examples, unfortunately, abound. Cold-water fish that populated these waters for hundreds of years now have to compete with warm water, non-native species. Previously clear, blue summer skies are too often tainted by eye-stinging, throat coughing smoke. The beauty of local lakes obscured for up to weeks by an ugly gray haze. Our own slice of heaven — summer outdoors — has become, at times, an indoor prison.

We cringe at such changes to our home. "What's happening here?" we ask ourselves, briefly, afraid of the answer. Many of us know, or suspect, it's global warming, but we don't know what to do about such a big problem. Others, already fearful, wince when they think: "When is the smoke going to return?" Anticipatory anxiety, it's called.

While few North Idaho residents have had to relocate due to extreme weather events, multiple years of wildfire smoke, recent heat waves and elevated summer temperatures have raised anxieties.

I heard that anxiety in the voice of a fifties-something nurse who took my blood pressure and pulse prior to a doctor visit on Sept. 1, 2022. Making small talk as she gathered the information, she asked:

"Did you go to the Fair?"

"No, I'm not a Fair person," I answered. "It's been too hot, too."

"Me neither," she said. "I've lived here my whole life. *It's never been like this before.*"

The angst of living in a town she no longer recognizes was palpable.

CHAPTER 4

Ancestral Lands

WRITING A BOOK ON local climate change needs to include this area's original inhabitants. This is, after all, their land, illegally and ruthlessly taken from them.

The Lake City was once part of the five-million-acre territory occupied by the Coeur d'Alene Tribe, in Eastern Washington, Northern Idaho and Western Montana. The Coeur d'Alenes were semi-nomadic and referred to themselves as Schitsu'umsh, meaning "Those who were found here" or "The discovered People" (CdATribe.com>History).

This has been the tribe's home for at least 12,000 years, according to Caj Matheson, Natural Resources Director for the Tribe, in a 2019 interview with the author. The Tribe's website says the Coeur d'Alenes are "the sum of uncounted centuries of untold generations."

The following highlights the marked difference in worldviews between the Indigenous and white cultures:

"What you people call your natural resources our people call our relatives."
Oren Lyons,
faith keeper of the Onondaga tribe

My worldview as a white man, of which I'd hardly had a passing thought, was completely upended when I first read the above quotation nearly twenty years ago. I couldn't imagine being related to a tree, water or fertile land as a relative. While it seemed somehow preposterous, it was at the same time compelling. I wanted to know more, so I arranged an interview with Matheson.

During the interview, he referred to animals as *he* or *she*, not *it*. They're not just fish, but fish or water people. He shared that he'd walked and picked huckleberries where his great-great-grandmother had, 150 years before. His love for her reminded me of my love for my grandfather.

Matheson enjoyed sharing the storytelling he grew up with and continues with his own children. Coyote stories are at their heart, thousands of years old. They highlight long-standing tribal values, important in raising children and preserving their heritage. The children learn about the ways of the external, natural world through listening to the stories. Not every Tribal child, however, hears Coyote stories, for various reasons.

Coyote stories are generally told during the winter, when the family is together and inside more, Matheson said. In the stories, Coyote is cunning, a trickster and deceiving (like the cartoon character Wile E. Coyote years ago). He's also a source of comic relief when he messes things up, most often when trying to do things for himself, rather than for the community, or the greater good, Matheson related. With human traits in the stories, Coyote and other animals are referred to as *he* and *she*.

Matheson explained that the animals have a right to be alive. The Tribe, after all, lives in their backyard, so they need to have a "give and take" with all plants and animals. Out of respect and the desire to live in a harmonious way with all of life, the Tribe should never take more than it needs from them, Matheson emphasized.

I finally began to understand, albeit through a privileged white man's eyes, what was meant by "our relatives."

In late July 2023, I met with the Tribe's Climate Change Resilience Coordinator, Aiyana James. A recent Oregon State University graduate, James represents a younger generation in the Tribe's climate change efforts.

Asked about the Tribe's biggest climate change issues, James described salmon restoration, protecting Coeur d'Alene Lake, and wetlands restoration. Since that interview the Tribe has completed a Climate Impacts Assessment, which they presented to the Tribal Council. The main focus of their work now includes: 1) reducing carbon emissions; 2) food sovereignty; 3) energy sovereignty; 4) water quality and quantity; and 5) zero waste.

James then shared insights about the Tribe's Seasonal Calendar. Likely developed and passed down through many generations, it shows when plants and animals are available to be picked or harvested. For many generations, huckleberries, for example, have been available to be picked in mid-August. But James had to pick them three weeks early in 2023 (the weekend of July 20th), so they wouldn't waste away.

It's a major concern when the Seasonal Calendar can no longer be relied on, due to the shifting climate. An important orientation for the Tribe would be lost.

In James' view, the source of climate change is the disconnect between humans and nature, and the resulting lack of care for nature. That needs to be "built back" through a greater sense of community and empathy, she emphasized.

James noted that Indigenous people are "the closest people to this planet." Therefore, their knowledge of the land needs to be acknowledged, and they must be allowed to work with it in their own way.

CHAPTER 5

Coeur d'Alene's Climate History

TO BRING TO LIFE Coeur d'Alene's climate history, the town's annual average temperature will be displayed over periods of time, memories of longtime residents who lived through those times shared and newspaper accounts of the events described.

Figure 3 (below) shows the average yearly temperatures for Coeur d'Alene, from 1893, when the weather was first recorded, to the year 2022 (the latest data). Weather data is taken at the Coeur d'Alene Airport; annual temperatures are averaged daily high and low temperatures. The trend line shows the temperature increase, from less than 47°F in 1893 to over 49°F in 2022.

We can also see how quickly or slowly temperatures have changed over time. Note the Trend Data at the bottom left of Figure 3. It shows a 0.22°F per decade temperature increase, from 1893 to 2022. Translated, that means it took forty-six years (1.0 divided by .22) for the temperature to increase one degree. The shorter the period of time it takes to increase the temperature one degree, the stronger the warming (and vice versa).

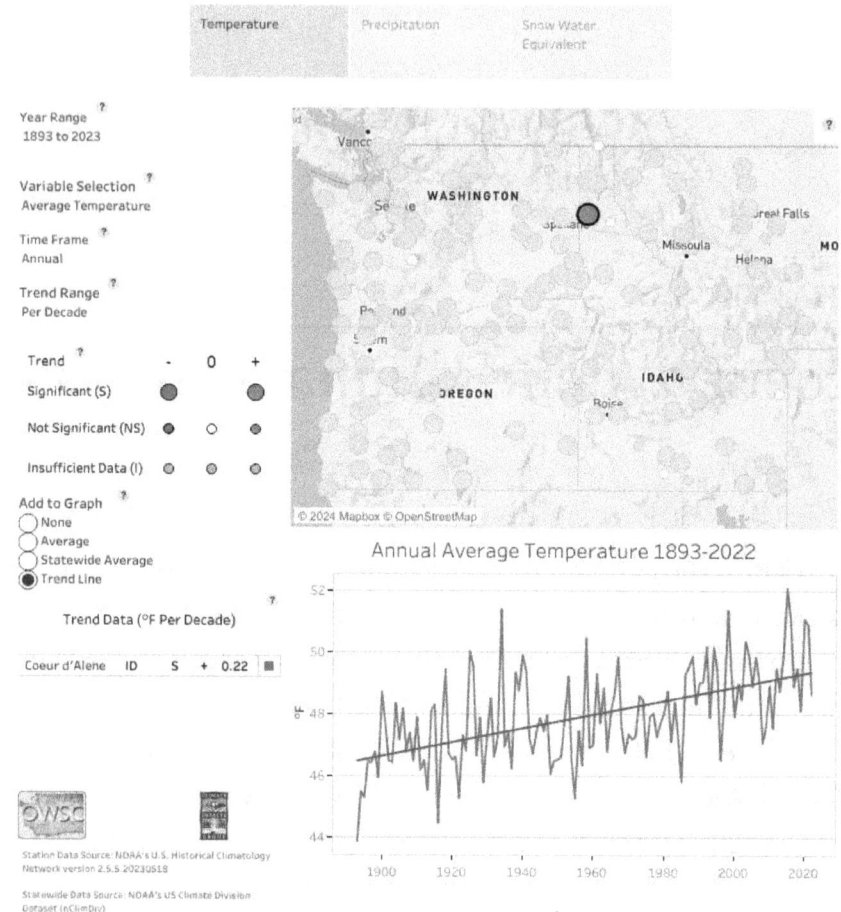

Figure 3
https://climate.washington.edu/climate-data/trendanalysisapp/

The 1910s and 1920s

The McLain family has been in Coeur d'Alene nearly as long as there's been a Coeur d' Alene. Sarah and Ellen, third-generation Coeur d'Alene natives, live here, while sister, Susan, and younger brother, Patrick, live elsewhere. Their mother, Betty, was one of my mother's dearest friends.

With a lively spirit, Sarah was generous in sharing her parents' memories of growing up in the Lake City, in addition to her own. She related a few stories regarding her maternal grandmother, Hazel Cardwell, in the early 1900s as well. Hazel played on the Coeur d'Alene High School's girls basketball team. One game she went by electric train to Ross Point and then by sleigh to Rathdrum (Dahlgren, D. & Kincaid, S.C., 2009, 85).

Born in Coeur d'Alene in 1891, Hazel (we called her "Goggy"), was a young teacher at a one-room schoolhouse in Squaw Bay (now Neachen Bay, on the lake), Sarah said. Hazel boarded with a ranching family and rode a horse to school since "it was too difficult to go home in the winter" due to all of the snow, Sarah recounted. Hazel needed to build fires for the students to keep warm as well.

Sarah heard that there was "lots of snow" during those times, but "they just dealt with it." Records indicate snowfall was above fifty inches per year for Hazel's first eight years, then below fifty inches for the next thirteen. When Hazel was twenty-one, more than fifty inches fell over the year, then amounts dropped again until she was twenty-five. A whopping 89.8 inches of snow fell that year, in 1916.

Sarah's mother, Betty, didn't talk much about the weather, or missing school because of it, Sarah said. Neither did either of my parents. We have little idea what it was like for people in those times to live in such cold weather. Newspaper accounts noted only the high and low temperatures for the day. The weather was simply not "news" back then, like it is today.

Figure 4 depicts the bountiful snow of those times. It's a sled dog race in downtown Coeur d'Alene, on Feb. 5, 1916. "Not less than 2,000 people lined Sherman Avenue, a larger crowd than ever assembled for a circus parade. And, the races themselves were better than Charlie Chaplin," the *Coeur d'Alene Evening Press* reported. The races ended in 1919, likely due to World War I and the significant number of deaths from the Spanish flu epidemic. After the race sponsor, insurance and realty agent Joe Petersen, died in 1933, a tribute/revival race was held in 1938 (*Coeur d'Alene Press*, 1938).

Figure 4

A bit earlier, about the year 1900, throngs of community members are shown ice skating on Coeur d'Alene Lake, off Tubbs Hill (Figure 5). Winters cold enough to freeze ice thick enough on which many could safely ice skate was clearly not a concern.

The late 1920s and 1930s brought very dry conditions. A drought from 1929-30 was the driest on record: less than thirty-three inches of moisture fell and only 17.8 inches of snow during that two-year period. After that, it continued extremely dry, from 1934-37, during the Dust Bowl (Mann, 2012).

Tubbs Hill circa 1900
Photo courtesy of the Museum of North Idaho

Figure 5

Temperature and Snowfall Extremes: 1916 to 1937

Coeur d'Alene's weather showed "extremes" from the late 1910s through the late 1930s, followed by less extreme fluctuations since. The year 1916, for example, registered the coldest year-round temperature in the Lake City's history, just over 43°F (Mann, 2012). For perspective, the average annual temperature in 2022 was nearly 49°F, almost six degrees warmer.

During 1916 temperatures were below average eight of the twelve months. The second half of January saw -14°F and -18°F readings. Along with those lower temperatures, snow amounts were higher than normal for the season. For example, March's total of 23.4 inches compared to the average of only 1.1 inches. For the entire year, 89.8 inches of new snow fell. The "normal" amount was only 29.2 inches (Mann, 2012).

It was the opposite in 1934. That year saw the third highest annual average temperature ever recorded in town. Only 1998 and

2015 have been hotter. Eleven of the twelve months were above average temperature. Not surprisingly, snow levels were abnormally low — half the normal amounts: 14.2 inches vs. 28.1 inches (Mann, 2012).

Just three years later — January 1937 — was the coldest January on record, averaging in minus digits: -2.29°F. There were nineteen sub-zero mornings (Mann, 2012). That winter my father milked cows in 45°F below zero weather in Rupert (south central Idaho) when he was twelve years old. Contrast that with the following summer in the Lake City, when it was the third hottest in recorded history.

It was a time of weather extremes indeed.

Through these times, from 1924 to 1938, Coeur d'Alene had an ice hockey team: the Eskimos. When games were played in town, they were generally on Fernan Lake, since it was the first to freeze. The Eskimos also played on the lake in front of the city park, and a temporary rink at Memorial Field. Prepared by the American Legion, it was 100' by 200,' banked with sawdust and filled with water to freeze.

Most games were played at two in the afternoon to provide enough light on the outdoor rinks. Practices on Fernan Lake were at night, the rink lit with oil lamps or by burning oil-soaked tires.

In 1927, the Coeur d'Alene Eskimos won the Inland Empire Championship and the coveted Davenport hockey trophy (Mitchell, 2022). Members of the 1930 team are shown in Figure 6.

Figure 6
(Photo courtesy of Museum of North Idaho)
Photo of the 1930 Eskimo team, taken at City Park.
Photo appeared in *The Spokesman-Review* on Feb. 20, 1930.

The Flood of 1933

The timber industry was vital to Coeur d'Alene's early growth. "In 1928, seven [lumber] mills operated in and around the city of Coeur d'Alene; by the end of the 1930s only three remained" (Singletary, 1933). What caused the mills' demise? The Great Depression, followed by the major flood of 1933.

Historian Robert Singletary described the 1933 flood. On December 16th, "light snow showers ... soon turned into a steady rain. The rain, accompanied by warm winds, resulted in the melting of the snowpack in the mountains." Within four days, the lake level rose to seven feet above normal. "The rain continued, and three days before Christmas, a violent storm swept over the area uprooting trees, downing telephone and electric lines and ripping off roofs."

Another two-foot rise in the lake and volunteers began piling rocks and sandbags along the city beach from First Street and around the Fort Grounds to the bridge at River Avenue. Floodwaters soon covered the grounds at the Blackwell mill, parts of Sherman Avenue and the area around the Milwaukee [train] depot, the Desert Hotel and the Electric Line depot (Singletary, 2018).

The tracks of the Great Northern Railroad were completely covered. Logs and trees floating down the lake made a "big" log jam in front of the Blackwell Bridge, threatening to take it out. Tugboats, one operated by Fred Murphy, and the use of dynamite to blow the trees apart, saved the bridge.

Newspaper accounts showed the flood's impact went far beyond our town: "The entire area from Coeur d'Alene [east] to Wallace was a 'vast inland sea.' All towns in the Coeur d'Alene district except Kellogg are cut off from the world, with efforts by the national guard to drop mail and food to the stricken communities." Hundreds were homeless in Wallace, with the mercury around zero (*The Spokesman-Review*, 1933).

Three years after the 1933 flood, a 1.5-mile flood wall and levee were built around the old Fort Sherman grounds. It was completed in 1940. Walking along the concrete abutments of the levee today, every eight feet are post holes for I-beams to be inserted, to construct an additional four-foot-high wall, if needed. To date, the wall hasn't been needed. It's tested every five years, Singletary mentioned, to make sure the planks and levee system are in working order.

COEUR D'ALENE'S CLIMATE HISTORY

The 1950s & 1960s

A number of the long-time residents interviewed mentioned the decade of the '50s, into the '60s. Gerry House was interviewed by *The Spokesman-Review* in 2007. House said his great-grandparents were among the first settlers in the area.

In the 1950s, House recalled, "people regularly drove across Hayden Lake, though not always without trouble. His uncle once put a hay truck through the ice near Clark Point. 'A lot of people would commute up and down the lake,' he said" (Hagengruber, 2007).

The ice on Coeur d'Alene Lake was thick enough to drive on as well, as noted in the 1950 photo (Figure 7). Two tractors are clearing snow off the lake for ice skating. Note the ice skater in the background.

Figure 7
(Photo courtesy of Museum of North Idaho)

I interviewed a number of local residents at the Lake City (formerly Senior) Center in early March 2019. Jeannie Edinger, then seventy-six, was sitting near her husband, waiting to get their taxes filed. A native of Coeur d'Alene since she was six years old, Mrs. Edinger remembered winters being much different than they are now:

> I feel like we're getting more Seattle weather [now]. The winters were a lot worse [in the 1940s and 1950s]. I remember building igloos, having snow like that.... I was probably about 10, maybe...Oh yeah, a lot more snow [then]. We would build them [igloos] into the snowbank at home. They wouldn't be called igloos, they were more like snow forts.... We had a gal in my class.
>
> Her dad used to drive her across the lake [for school, during winter]. I graduated in 1960, so it probably would've been in the '40s or '50s. He would take her most every day. I think her name was Murphy, Joan Murphy.

Joan Murphy's father was Fred Murphy.

In addition, Mrs. Edinger added: "There's a lot more rain now in the winters." Although Edinger expressed no concerns about the change in climate, she continued: "I do worry about the global warming a lot," especially as it would affect their two-year-old granddaughter's "lifestyle."

Mary Lou Reed is another long-time resident whose experience is noteworthy. An Idaho State Senator from 1984 to 1996 and instrumental in the founding of the Human Rights Education

Institute in town, Reed continues to participate in the Coeur d'Alene political scene.

She and husband Scott came to Coeur d'Alene in 1956 "in search of a four-season climate. We wanted snow to ski on and most important, ice on which to skate." For the 2015 holiday season, Reed wrote an article for *The Inlander* magazine about "local winters when they were longer and much colder than those we enjoy/endure today" (Reed, 2015).

The Reeds were not disappointed their first year here, in 1956. The temperature dropped to zero before Thanksgiving and they skated on both Avondale and Hayden Lakes: "It was late April before the ice finally melted off Hayden" (Reed, 2015).

Old-timers told them of the ice industry [Coeur d'Alene's Valley Ice & Fuel Co.] years before "would deliver large chunks of ice to the back porch in boxes of homes around town. The ice was carved out in massive chunks from Fernan Lake and Coeur d'Alene Lake and stored in a warehouse just a few blocks from Sanders Beach" (Reed, 2015).

That memory sparks my own. Shortly after my parents bought our family cabin on Lower Twin Lake in 1972, Mom met Merl Miley, the owner of Twin Echo Resort. Miley said he'd been carving out chunks of ice from the frozen lake for decades. He'd then store it in sawdust underground in a nearby building. The ice would be sold to cabin owners and visitors during the summer. Figure 13 (in chapter 8) shows Miley, with two friends, cutting out the lake ice.

Reed continued about ice skating: "Up until very recent times, Fernan Lake would freeze solid every year. Families would turn out en masse to skate, stumble, fall and just have a grand old time. Clamp-on skates, that could be adjusted to boot size, were passed down from sibling to sibling." But she's noticed, "It has been a long time since

Coeur d'Alene Lake has completely frozen over. Even the smaller, shallower lakes can't be relied upon for solid…ice" (Reed, 2015).

Reed then recalled Joan Murphy skating to school when ice covered Coeur d'Alene Lake in the '40s or '50s. While Edinger remembered Murphy being taken to school by her father, either ice skating together or on his snow machine, the frozen lake offered them both safe passage. Reed then added a fun tidbit to the story. Fred's son, Loren, told Reed about "the mystic joy of skating home by moonlight after a high school event" (Reed, 2015).

When will climate change hit?

Walloped: Winter of 1968-69

The majority of interviewee's reminiscences revolved around the 1968-1969 winter. It was epic. Stories from the *Coeur d'Alene Press* will describe what happened, followed by local residents' memories of those times.

The winter before my grandparents passed away was especially harsh. It began with an overnight low of -26°F on Dec. 30, 1968, the fourth lowest temperature in town since weather records were kept. Eight inches of new snow began falling on New Year's Eve, bringing the total to just over twenty-eight inches. January 1st was the snowiest since 1935 (*Coeur d'Alene Press*, Dec. 30, 1968).

A second "severe" storm hit a week later, closing schools. Seven to eight inches of fresh snow blanketed the Lake City. An additional nine inches fell over the weekend, toppling trees under the weight of the snow. A seventy-mile section of highway 95A, from Wolf Lodge to north of St. Maries, was closed (*Coeur d'Alene Press*, Jan. 13, 1969).

With a month of nearly non-stop plowing, hauling snow, sanding the city streets, street superintendent Reed Walker said: "This has been the worst conditions for plowing in the 14 years I have been with the department (*Coeur d'Alene Press*, Jan. 22, 1969). And it only got worse.

A weekend storm—"by far the most severe of the season"—dumped fourteen inches of new snow in the Lake City, more in nearby towns. The *Coeur d'Alene Press* reported: "Winds accompanying the storm blew it into insurmountable drifts and closed all highways and roads for a time Sunday" ("Area Still Staggering Under Heavy Snow and Winds," Jan. 27, 1969).

By Tuesday, Jan. 29th, only ten of the twenty school days in January had been held, due to cold temperatures, snow amounts and road conditions (*Coeur d'Alene Press*, Jan. 29, 1969). Snow drifts as high as sixteen feet were reported in some areas (*Coeur d'Alene Press*, Jan. 31, 1969).

A state of emergency in ten north Idaho counties, including Kootenai, was declared by Governor Don Samuelson on Jan. 31, 1969. Snow conditions were the most severe in a half century (*Coeur d'Alene Press*, Feb. 1, 1969). A total of 117.8 inches of snow fell during that winter, the fifth highest total in the city's history (Mann, Jan. 9, 2020).

Hayden's Rustler's Roost restaurant was the site of additional shared memories of longtime residents. I met a seventy-three-year-old Coeur d'Alene native there. That winter he remembered snow drifts so high he couldn't see the fence posts, driving along State Highway 95:

> It was in January of 1969, we had six foot of snow out there on the level. It snowed all the way through January, it seemed like every day… At the end of January, I had to be out of town. At the end of January, we got a blizzard,

and those berms along the road, some of them were twelve feet high. That blizzard just blew them clear full. I got into Spokane, and it took me three days before I could get home because it took them [road crews] that long to get the roads open. And that was just another storm.

He continued:

Everybody was told to tie ribbons on your [radio] antennas on your cars here in town because snow was piled up, they couldn't handle it." (See Figure 8) ... [The road crews] plowed the snow [from what was then Highway 10, now Sherman Ave., in downtown Coeur d' Alene] into the middle, and a machine came along and picked it up and put it in dump trucks and hauled it away. There was no place in Coeur d'Alene to put it.... That was in the '50s and '60s.

Next, on a deeply gray day in April 2019, Louise Oliver was impatiently waiting to be interviewed at the Lake City Center. She made it known she was miffed I hadn't talked to her already. Louise and her husband came from Albuquerque, New Mexico in 1966 for his job here, she said. For forty years they had a 200-acre ranch in Cougar Gulch, south of town.

Eighty-eight years young, Oliver quickly remembered the foolishness of their twelve-year-old son at the time. He jumped into the snow and got buried in it, upside down. He then had to get pulled out. Snow was so deep, it went over the fence post on their property, she said. The cows walked over the posts, behind the farm house, and got lost as well. Oliver and her sons had to herd them back where they belonged.

Figure 8
Coeur d'Alene Press photo, January 22, 1969

Street Supt. Reed Walker points to a flag tied to an automobile's antenna during the 1968-69 snowstorms. The flag allowed drivers to be seen over the snow piled between the traffic lanes.

Yes, she'd seen people drive on the lake when it was frozen. Also, there was "so much snow" the road crews had to dump it into trucks and then into the lake during that 1968-1969 winter. When asked about more recent conditions, she expressed concern that "we are making" the weather change, which is "destroying our way of life." Oliver longed for the "much simpler life" of the 1940's and 1950's.

During that infamous winter, Sarah McLain's father, Mac, walked across the frozen lake to their cabin with Sarah, younger brother Patrick and two neighbor friends. Later, Sarah heard that her mother got mad at Mac for "taking that risk with other people's kids." However, Sarah said she wasn't afraid or worried at the time. It was a "high adventure," walking on the ice with long wooden poles.

Remembering an announcement in the 1969 paper the next day, that anyone caught on the ice would be fined, brought hearty laughter from McLain. Neither Sarah nor husband Tom Hughsby could remember the lake freezing over since the winter of 1968-69.

She remembered missing "an entire month" (January) of school during that winter. Similar to the newspaper account, she said the snow started on New Year's Eve (during Christmas break) and it "just kept coming and coming."

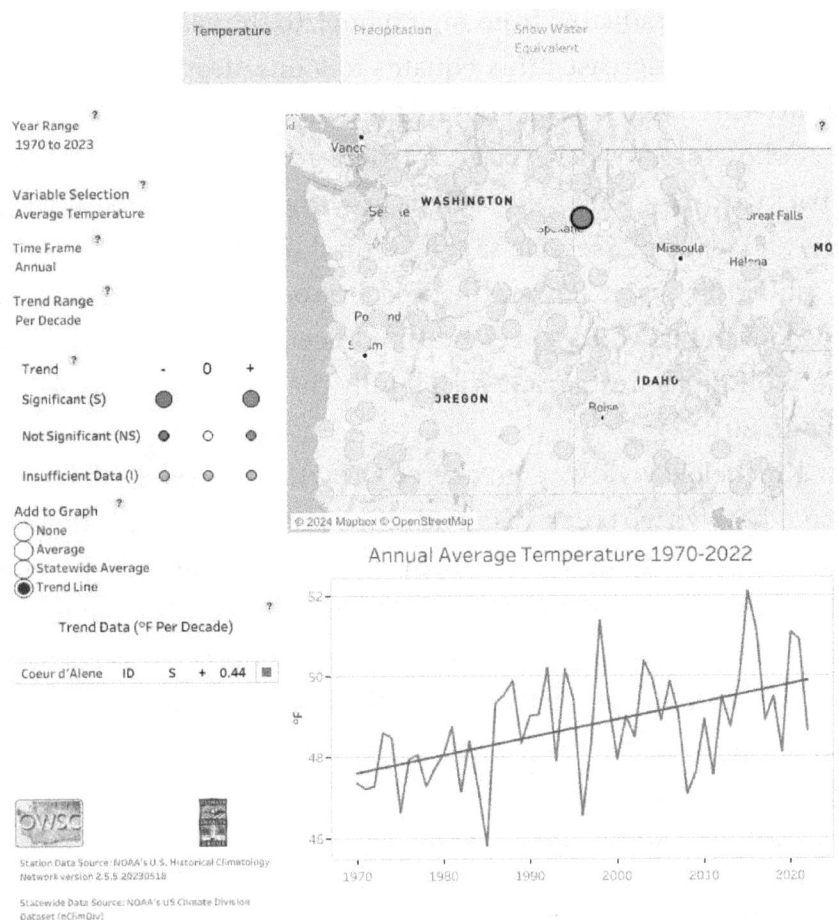

Figure 9

https://climate.washington.edu/climate-data/trendanalysisapp/

After their next-door neighbor's dad made a step pathway of snow up their garage, Sarah and the neighbors hauled inner tubes up and slid back down to their backyard, taking pictures of each other with Kodak cameras, she remembered. Sarah indicated that "we always had snow to make tunnels in, to play in, which is different now." Figure 9 shows how quickly the temperature rose, from

1970, when I graduated from high school, to the year 2022 — a .44° F per decade increase. This equates to a one-degree F. rise every twenty-three years, a time of rapid warming.

Following our visit, Sarah introduced me to Ray Kinchelow, a backyard neighbor with a light gray mustache, tufts of white hair and black glasses. He said he retired in 1992 as a lineman for Washington Water Power, the predecessor to Avista Utilities. He was looking forward to celebrating his eighty-ninth birthday at the Wolf Lodge restaurant with family the day after we spoke, on May 3, 2019.

Kinchelow vividly remembered the winter of 1968-69. He was home barely two weeks the entire winter, dealing with downed power lines and outages in Coeur d'Alene and Casco Bay. When his employer heard livestock were starving in Mica Bay that winter, due to the owners being snowed in, Ray and his co-workers delivered hay bales for the cattle. When the residents came out of their homes, their faces were black—they'd had to burn indoor fires to keep warm in the freezing cold.

Climate change approaches.

The 1970s to the 1990s

Melvin Rettman, my former wife Lisa's now deceased father, was eighty-seven when he offered his memories to Lisa in Minnesota. Mel lived in Harrison, south of the Lake City, and Coeur d'Alene in 1972-1973 with his wife and children. He'd fished in Fernan Lake as early as 1970 as well, he added. The fish were "biting good" during both the ice fishing and lake fishing in the 1970s.

But by the early 1990s, when he came to visit Lisa in Hayden, "the fishing was way, way down," he said. Why? Mel checked the *Farmer's Almanac*, his authority on such matters. The *Almanac* said it was because the water had gotten warmer. That was good enough for him.

When he and the family lived in Harrison, he remembers logging trucks hauling the fallen trees into Coeur d'Alene Lake year-round. Skidders would pull the fallen trees out of the forests, across the winter snow, and dump them in the lake. There, tugboats would corral them for the ice-breaking trip across the frozen lake to the sawmills in the Lake City. The weather was cold enough to consistently freeze the lake solid.

Next, I interviewed Jim Kimball, an environmental engineer who un-retired in 2019. A longtime climate change activist, he offered a few observations regarding the weather in the 1970's.

The year 1973 was an "extreme drought" year and a warm summer in Coeur d'Alene, he said. The environmental engineer for the Idaho Department of Health & Welfare at the time, Jim discovered too high chloroform counts for recreational swimming at Honeysuckle Beach, on Hayden Lake. He had to post "No Swimming" signs along the beach. That year Twin Lakes also had an outbreak of "swimmer's itch" due to too warm water conditions, he recalled.

The heat and drought also affected the price of hay for his horse. Ordinarily $17.50 per ton, it jumped to $100 per ton, due to reduced yields of dryland hay fields. This required him to buy beet waste and pellets to feed his horse.

Ray Kinchelow, mentioned earlier, recalled Coeur d'Alene Lake freezing over during the winter of 1968-69, again in 1971 or 1972, and then a third time he couldn't recall when. The temperature dipped to 35° F below zero sometime in the 1970s as well, he thought. When he and his wife moved from Spokane to Coeur d'Alene in 1970, he said every winter "there was quite a bit of snow."

Things started changing in the 1980s, though, Kinchelow noted. Winters became "milder and milder," and there were "big snows few and far between." In addition, rain in winter became more common. Besides the cold weather and late snow in February 2019, he said he hadn't seen difficult winters since that time.

Kinchelow said he didn't plant flowers before Memorial Day in the past but now, this first weekend in May, he might get out to "dig up the soil." Asked about the frequency or severity of droughts, Ray said he now "waters a lot more than I used to."

Kinchelow has nine grandchildren and two great-grandchildren. Yes, he's concerned for them about these changes in the climate, mostly his seven-year-old great-granddaughter and two-year-old great-grandson. Declining snowpack means less water for their use, along with hotter and unhealthier conditions. He's most bothered, though, with the "factions" of people who are in denial about global warming.

Also noticing the milder, less snow-filled winters is Wes Hanson. A well-spoken and astute observer of the land, I interviewed him during the drought, high heat and wildfire smoke of 2015. He's lived on the Carter Farm, two miles south of Coeur d'Alene, for the past forty-seven years. Since 1997 the 100-acre farm has been protected by a conservation easement.

The winters were "pretty severe" when Hanson first arrived from Michigan about 1970, he said, with cold temperatures and plenty of snow. But "over the years they have tapered off," with less frequent snowfall and snow that melts off more quickly now. He added that winters are "noticeably more constricted," and with a later onset.

Less snow, which melts more quickly, has consequences he's noticed while walking the woods outside his home. Soil moisture has been greatly reduced, which has retarded the growth of white pines and fueled recent wildfires, he said. In addition, these changes have consequences on the identity of Coeur d'Alene, one tied to its beauty and comfortable climate.

Hunting and Fishing Impacts

I was fortunate to interview Denny Webster, who's familiar with the local hunting and fishing scene. He and his wife have lived in the Coeur d'Alene area since 1978. A retired Coeur d'Alene School District administrator, he offered his observations on a beautiful day in early May 2019 when I visited him at his home.

In the early 1980s, he said, hunting season for elk ended around Nov. 3rd. He needed a snowmobile to hunt them in deep snow in the backcountry. But now, he hasn't needed a snowmobile for years.

To highlight this point, he recounted a recent conversation with a "big time" hunter friend, Dan. He asked Dan: "Remember when we used to hunt in the Joe (St. Joe River area), in the early '80's, we rode snowmobiles all of the time? We never do that now, not for years." Dan said he hadn't thought about it, but Webster was right. He doesn't even own a snowmobile anymore — it's no longer necessary for the hunt.

The change, however, has been "so subtle, if you don't stop to reflect back," Webster noted, "you don't even notice the change."

He makes an important point, similar to the story of the frog in a slowly boiling pot not noticing the temperature rising. It's important to note that while humans are similar, they have the ability to act on threatening changes before it's too late. Hopefully we'll do so.

Webster then discussed the changes in big game hunting due to local climate change. When he first started to hunt deer in the '80s, he could get "quality" bucks during the rut, in the second week of November: "It was prime time to take off from work, to hunt deer."

The rut is when male deer pursue females who are ready to breed. Over the years, though, the rut is now probably two weeks later, in the Thanksgiving time period, for elk as well.

Archery season to hunt elk used to be the month of September, he said. Rifle season would start about Oct. 10th. When Webster bugled bucks (start on a high note, and slowly release it with your teeth, to make a mewing or chirping sound), he wouldn't get much of a response from the bucks. But now, with the rut later, into the rifle season, the elk respond.

Also, when Webster first arrived here in 1978 it wasn't uncommon for Coeur d'Alene Lake to freeze. Now he said it hasn't frozen over in many years. Lake Pend Oreille used to freeze "straight across" in December in the '80s as well. Now it hasn't come close in years. People used to be able to ride vehicles "back and forth" across Spirit Lake, too, but now it's rare.

Webster said he's ice fished on Priest Lake before, but in 2019, he was only able to do it for a few weeks, in late January, due to warmer winter temperatures. The ice was too unstable to walk on. Now he usually uses a boat. On smaller lakes, like Twin Lakes and Avondale, he could ice fish all the time during the winter. However, now he has to ask himself, year to year, "Can I this year?"

When Webster began fishing on rivers in 1979, in Montana and Idaho, he could get fish "pretty much" year-round, with no regulations. Now in Washington its "Hoot Owl" regulations don't allow fish to be caught after twelve noon due to water temperatures getting too warm. The trout get too stressed to bring them in, then release them. They may not survive the ordeal.

With higher water temperatures in the afternoon, people need to be much more "conscious" of the health of the fish, something they didn't need to do in the past, he noted.

COEUR D'ALENE'S CLIMATE HISTORY

Although humans have a hard time sensing a one- or two-degree temperature increase, as noted earlier, a one-degree temperature rise in the water for fish is "very big," Webster said. When water temperatures reach fifty degrees, for example, spiny ray fish begin to move in, and spawning, fish breeding and migration change, he emphasized.

Nowadays in the summer, the water gets "pretty skinny" (fisherman talk for shallow water), Webster said. More frequent low flows in the late summer now make it "very difficult" to get down the river to fish, something that has him "very concerned."

Webster noted rising river temperatures are also an issue. In the past, a canopy of trees provided shade, which helped to cool the rivers. But with fewer trees near the rivers now, water temperatures have risen. This threatens the survival of cold-water fish species.

Many of today's outdoor activities require Webster to keep a daily eye on both water levels and water temperatures. He didn't need to do either when he and his wife first moved here.

Gardening and Fishing Casualties

Following a January 2019 Audubon Society meeting at a local church, I was introduced to Midge. She didn't offer her last name. Later, her email responses to my questions came from MB.

Midge offered her own and her father's experiences in how growing vegetables and fishing have changed in the Lake City. In the past, Midge noted: "Lettuce could be planted the end of May. Now I have had to find a heat tolerant lettuce or it will be cooked by the sun. Green beans used to be wonderful. Now they are tough and the vines are not lush like they were. They dry out, especially pole beans.... *Beans, like cool nights: we don't have them anymore.* [author's emphasis]

"You were always lucky if you got many tomatoes due to cold nights. This last year (2018) we had a week of over 90 degrees so

57

early in the year that it cooked the blossoms off the tomatoes planted outside.... I have completely changed my type of tomatoes that I grow now....

"The transparent apples no longer do well here—they mature too quickly and almost cook on the tree. (Transparent apples are a top choice for apple pies, google says).

"Dad used to go fishing. I remember him saying the fish were not as active because the summers were getting too hot and that their flesh was not as 'tasty'; it was soft and mealy and not flaky" (MB email, Feb. 25, 2019).

1970s to the 1990s and beyond

An older gentleman arrived at the Lake City Center, with his daughter, a bit out of breath. He walked with a cane, white hair combed back, glasses perched on his nose. He had a happy smile that reminded me of my kindly grandfather. It was easy to begin to chat with T.J. Cornelsen as he sat down next to me in the open area of the center.

Then ninety-two-years old, he said he's lived in Coeur d'Alene since May 8, 1981, a date he quickly remembered. He "loves" living here, he said, due to "the four seasons" and "enjoying the snow."

Cornelsen proudly stated that he has an All-Terrain-Vehicle with a fifty-two-inch blade to clear the snow. He clears his own triple driveway, his neighbor's four-car driveway, plus two other neighbors' three-car driveway and two-car driveways. In addition, some twelve blocks of sidewalk on each side. Cornelsen then added the obvious: "I really like to get out in the snow."

About the local weather changes, Cornelsen continued: "It's a lot warmer than what it was."

"Back in the '80's, it'd be twenty-two degrees below. In '92 I worked for the Coeur d'Alene Sanitation Dept., working off part of my property tax. The temperature then was sixty-five degrees below zero, wind chill factor."

I then offered my father's story of milking cows in 45°F. below zero in Rupert, in the 1930s, with no wind chill factor.

Any other changes you've noticed in the weather, I asked.

"'96-'97, we had an ice storm, and we had snow and ice up until the latter part of March. I had three foot of snow in my backyard that year. Since then, it's been kind of up and down. Definitely getting warmer. This year (2019) the snow they had in February was unusual snow." It came on so late, too, he agreed—hardly any in December or January.

T.J. said he didn't think it's gotten hotter here in the summer: "If we get to 100 degrees, it won't stay that way for very long." The biggest thing was not as much snow since 1981, and it's not as cold as it used to be.

Asked if he had any idea why those changes have happened, he said: "I think it's due to global warming. I was up in Canada in '76 at the Columbia Ice Fields and have been there a few other times. The ice fields have receded about a mile and a half. You used to be able to walk to the ice fields and now you have to take transportation out there."

After his last comment, T.J. bounded up (at ninety-two!) and walked briskly into the Great Room, to play pinochle with his daughter and others.

In the center's pool room, Albert Neff, sixty-eight, got his pool cue out to play. Introduced by a fellow pool shark, Albert said he's lived in the Lake City-Post Falls area for the past fifty years. He said the seasons have changed over the years, and volunteered his view on global warming:

> What I remember is that the seasons were pretty much regular. Every year you knew just about when the winter

was going to start, about how much snow you were going to get, when it was going to go away. All the seasons were about the same as far as the length of the seasons. But now, it's completely different. The seasons are much more mild than when I was younger, growing up here. They've changed a lot. They're much more mild. It just seems like every year that goes by, it just seems like it gets more mild. The amount of snow that we got, that has changed drastically. I think if you talk with most people, especially natives, they will say that the seasons have changed. There's only one thing you can attribute that to: that's global warming. A lot of people will challenge that but I don't think scientists are crazy. There's something about global warming.

As a child, the summers were lush and green at the McLain cabin and on Tubbs Hill, Sarah remembered, with more moss and greenery until August. Now, "by June, it's looking dry already," she lamented, fueling "all the fires we've been having."

Not only is there more rain now vs. snow, she added, but "it feels like, too, we've been having heavier rain, just downpours now," plus "milder winters."

About summer: "I think it still cools down around in the summer, but we have longer spells of hotter weather, with 100-plus temperatures."

She hasn't noticed flowers blooming earlier but, "I sure have weeds in my yard that weren't there twenty years ago." The University of Idaho Extension Office told her they were "new" here. There's also much more weed growth than in the past.

McLain noted there's more drought now, and when it rains, there are more "long downpours" for days. In the past, there were more intermittent flows of water and heat. She mentioned as well the loss of another relic from the past: local ice sculpture festivals, last seen here about twenty-five years ago.

Not everyone I interviewed has seen a local impact from climate change. An anonymous female patron of the Lake City Center, for instance. Having lived in the Silver Valley and Coeur d'Alene area for "probably 60 some years," she said she hadn't seen much difference in the weather over that time span. Except, "It seems that the length of time that we have a hot spell lasts a little longer now than it used to. Where it used to be two or three days, now it'll be a week, and a bit hotter sometimes, but not necessarily."

She noted no difference in amounts of winter snow or the snow vs. rain mix, or any noticeable difference in temperatures during the winter: "I'd say it's about the same as far as temperatures in the winter. Not that much more cold in history when they've showed it on TV than we had in the past." "Flowers aren't blooming earlier either," she said, in response to the interviewer's question.

Asked if she had any idea why it's stayed hotter longer, she replied: "I do think we are experiencing global warming, but I couldn't point my finger at that and say that's what it is. I've seen quite a bit of scientific evidence from when I watch different television shows. But not changes here. I haven't personally seen anything that I could definitely point my finger at, and say that's definitely because of global warming."

Ice Storm '96

On November 19, 1996 a massive ice storm hit North Idaho and the greater Spokane area.

"Tens of thousands" of tree limbs fell on power lines, causing an estimated 10,000 households across Kootenai County to be without power. Nearly the entire county was dark. All public schools were closed for the day (*Coeur d'Alene Press*, Nov. 20, 1996). At least 100,000 homes and businesses in the Coeur d'Alene and Spokane area lost power (*Coeur d'Alene Press*, Nov. 24, 1996). It wasn't restored to all customers for two weeks, until Dec. 2nd.

Even before the ice storm was over, on Nov. 26, the operations manager for Kootenai Electric Cooperative (KEC), Gene Pope, declared it had "caused the worst damage to the Kootenai Electric system that we have ever seen…The damage is even worse than that caused by Firestorm '91, and even the winter of 1968-69" (*Coeur d'Alene Press*, Dec. 6, 1996).

At the east end of Coeur d'Alene, I walked a few blocks after the storm to check on Mom and Dad. Fortunately, they had their power already restored. I wasn't so lucky. When it was time to sleep, I was scared when I couldn't get warm, even with every blanket I had on top of me. After that I slept at my parent's place for a few days until my power was restored. Others in the Spokane area were not so fortunate. Some were without power for two weeks, I recall.

Record Snowfall

The winter of 2007-'08 was the snowiest in the Lake City's history, with 127 inches falling by Feb. 6, 2008. Coeur d'Alene students hadn't been excused from school because of snow in at least six years but that winter they had three snow days (Cuniff, *The Spokesman-Review*, 2008).

Windstorm of January 2021

A severe windstorm in January 2021 caused "staggering" damage to Kootenai Electric Cooperative's system, which provides power to rural areas of the county. According to a 2023 KEC newsletter:

At its height, more than half of KEC's members were without power. The storm caused nearly $2 million in damages...After the storm, a federal emergency was declared for Kootenai County which allowed KEC to apply for ... mitigation grant[s] from the Federal Emergency Management Agency (FEMA).

Subsequently, some overhead lines along Twin Lakes and twenty-two miles in a Spirit Lake subdivision have been converted to underground (Kootenai Power Cooperative Power Lines newsletter, 2023).

CHAPTER 6

Latest Climate Casualties

Rising Summer Temperatures

IN A MARCH 2020 column in the *Coeur d'Alene Press*, weatherman Randy Mann depicted dramatic summer warming since the 1990s (Figure 10). Mann used more reliable Spokane International Airport temperature data rather than from Coeur d'Alene. During the months of June, July and August, "daily average readings of the high and low have warmed up nearly 3 degrees Fahrenheit since the 1990s.

"For the decade of the 2010s, the airport reported a reading of 68.76 degrees. In the 1990s, it was only 66.08 degrees." Mann continued: "In terms of the annual average [temperature] at the airport, the last decade has seen the greatest warming, with an average reading of 49 degrees Fahrenheit. In the decades from the 1950s through the 2000s, average temperatures were 46.91 degrees during the chilly decade of the 1970s, to 47.89 degrees in the 1990s."

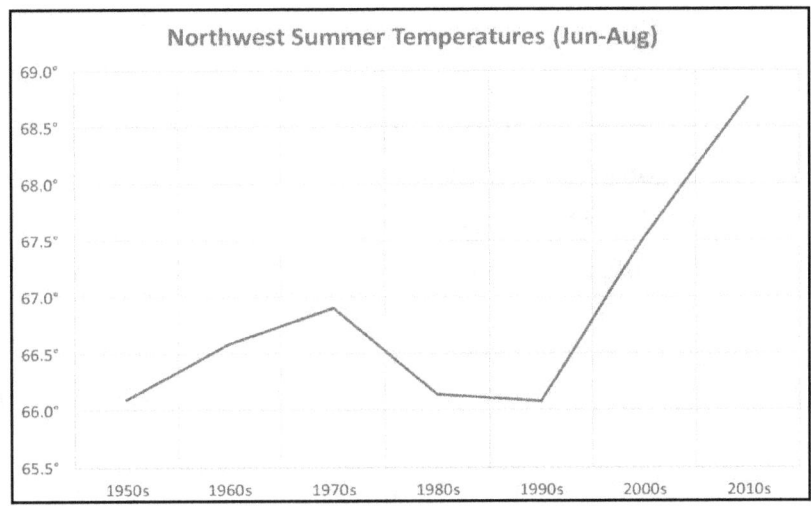

Figure 10
(Courtesy of Randy Mann)

Harbinger of the Future: The Year 2015

My former wife, Lisa, and I weren't prepared for the heat that hit us during the summer of 2015. A small air conditioner we bought was too small to cool the kitchen and living room. To attempt to stay cool, we ran a living room fan by ten a.m. and portable fans in the bedrooms all night. Without air conditioning, sleeping was miserable. Later, in August, Lisa's allergies worsened during heavy wildfire smoke. She struggled to breathe, at one point fearing she was dying.

The blistering hot weather was preceded by a winter like no other. An article in *The Spokesman-Review* called it, the Northwest's "trial run for future conditions under climate change…" A new term had to be coined to describe it: "snow drought." Not drought as we know it, but "near-normal precipitation, [with] temperatures about four degrees warmer than normal [that] kept snow from piling up in the mountains. The snowpacks that did form, melted off early" (Kramer, B., 2015).

Climate scientists issued a warning: "By around 2040, many ski areas will struggle with winter precipitation that falls as rain instead of snow…and even rainy areas will need to conserve water…We're seeing conditions this year [2015] that will be average by 2040 or later," said Phillip Mote, director of the Oregon Climate Change Research Institute at Oregon State University (Kramer, B., 2015).

The snow drought crippled the 2014-15 ski season. For only the second time in its history, Mt. Spokane ski resort suspended ski operations in mid-Feb. 2015 due to lack of snow (Clark, R., gonzagabulletin.com). No snow fell in February. Temperatures in the fifties, and rain, made it impossible to make snow, even if they'd had the equipment to do so, which they didn't. It was Mt. Spokane's second shortest season in the past 25 years (Clark, R., gonzagabulletin.com). Silver Mountain Ski Resort, near Kellogg, Idaho, also closed early.

Snowfall in the Lake City started off very slowly as well. By Dec. 19, 2014, only 2.3 inches of snow had fallen at meteorologist Cliff Harris' Coeur d'Alene home. Normal by that time would have been twenty inches (Harris, C., 2014, Dec. 22). The first big winter snow wasn't until Dec. 27th. Nearly a foot fell, along with more normal, colder temperatures (Harris, C., 2014, Dec. 29).

But by mid-January 2015, a "whopping" eighty-three percent of precipitation fell as rain. A typical winter would have sixty percent rain or mixed precipitation, and less than forty percent snow, according to Harris. A "big dump" of snow (four to six inches), for example, on Jan. 4th was "extremely wet," with temperatures hovering near the freezing point (Harris, C., 2015).

Nine months in 2015 were warmer than normal in Coeur d'Alene. Three months were substantially warmer than normal: June (9.5°F), October (6.2°F) and February (5.3°F).

That June it was 105°F on the 28th, the hottest June day in Coeur d'Alene's recorded history (until it was broken in 2021). It was also race day of the Ironman Coeur d'Alene, the hottest by a whopping 13°F (Cousins, K., 2015). Ironman medical tent coordinator Stan Foster said 535 people were taken to the medical tent when temperatures rose into the 100s that day. Five people had to go to the hospital, he told KREM-2 (Greene, B., 2021).

Seven days in July were over 90°F and eight more in August. At local meteorologist Harris' Lake City home, sixteen days in a row had high temperatures at or above 90°F, from June 25th to July 10th. On July 8th a Wildfire Smoke Advisory was issued for the Idaho Panhandle as well, due to fires in Idaho, Washington, and British Columbia. Smoke was everywhere.

In the face of the unrelenting heat, local gardeners were cautioned by Elaine Cerny, in her July 19, 2015 *Coeur d'Alene Press* column:

> I hope all gardeners have managed to keep your plants alive. With all the unprecedented heat, day after day, it hasn't been easy. Even the supposed 'heat lovers,' such as tomatoes, have been affected as they do not set new fruit when the temperatures stay above 90 degrees. They may bloom, but then the flowers just fall off. Most of the perennials have been blooming, but they are weeks ahead of a normal year. Those that normally bloom in late July were done by the first of the month" (Cerny, E., 2015). [author's emphasis]

It wasn't cool overnight either. July saw twelve nights well over 60°F. minimum temperature, then nine nights in August surpassed that overnight low.

Very low stream flows from the anemic winter snowpack, and too warm lake and river temperatures caused widespread fish die-offs across the Pacific Northwest (Fourth National Climate Assessment, 9). This included hundreds of thousands of sockeye salmon in the Columbia and Snake River basins (Fourth National Climate Assessment, 9). Local Fish & Wildlife Dept. officials denied any such fish die-offs in local rivers or lakes.

Lack of rainfall, along with wildfires in Washington, made the situation even worse. By mid-August, the Lake City's fire danger rose to the extreme level. On August 18th Tubbs Hill was hit with its fourteenth fire of the year, twice the normal number (Selle, 2015).

The lack of rainfall, along with two days of high winds in mid-month, caused climatologist Harris to raise these serious concerns: *"I've been keeping the weather here for over a quarter of a century, and I've never seen such dangerous levels for fire and conditions like we have now."* [author's emphasis] He added: "One spark, and we could see whole hillsides go up in smoke—It's a very dangerous situation, and we all need to be aware of it (Cousins, 2015).

Fires and smoke dominated the headlines. By the end of August, emergency closures of all National Forest System and Bureau of Land Management lands were ordered due to wildfires (Pearson, 2015). The forests were a tinderbox.

The Idaho Panhandle National Forest experienced *more forest fires [in 2015] than the last forty-six years combined* (51,000 acres, as of Sept. 2015), according to a publication of Kootenai Environmental Alliance.

"'It is unprecedented,' said Shoshana Cooper, acting public affairs officer for the Panhandle National Forest. 'We haven't had this much fire in the forest since 1926'" (Parrish, J., 2015).

The year also brought the longest Harmful Algal Bloom (HAB) on record (aka toxic blue-green algae) in Idaho—167 days, from June to December on Fernan Lake. Hayden Lake had an eighty-nine-day long HAB. Avondale, Hauser and Cocolalla lakes as well had HABs, as did nine other bodies of water across the state, according to the Idaho Department of Environmental Quality website.

Of the 171 summers since 1895, how did the summer of 2015 stack up in terms of overall temperature? Fourth hottest, behind 1967, 1961 and 1938. Thirty-nine of the ninety days of summer had high temperatures 90°F or above (Mann, Jan. 11, 2016).

The summer of 2017 wasn't much cooler. Thirty-four days were at or above 90°F. It was dry and smoky as well. Only 1.31 inches of rain fell between June 1 and August 31, putting 2017 in the top five for driest summers here in recorded history (Mann, 2017, Sept. 18).

Six days that summer Spokane's Air Quality Index (AQI) registered as unhealthy for everyone. Three more days were very unhealthy (Vestal, S., 2019). Local wildfire smoke was so heavy, according to a Lake City High School football player, outdoor practices had to be held indoors and a few early season games canceled (Hanna, R., June 2022 interview).

The air quality was unhealthy in 2018 as well. Ten days in Spokane it was unhealthy for everyone; two days were very unhealthy (Vestal, 2019).

Wes Hanson spoke eloquently about what the loss of snow means for our home and our place in it. "Without consistently heavy snows, we're missing the sponge capacity that winter provides to absorb water, hold it and then gradually release it." He noted a "sorrow" in all of us, "not only over losing the richness of the land around us because of drought and changes to the climate, but we're losing what's defined us as a community."

"The reason people come here," he added, "is the beauty of this place, which is intimately tied to the water and landscape, and the proximity to nature.... Climate change is altering nature."

In those few words, Hanson captured the major theme of this book. Now to more recent times.

The 2021 Heat Wave

The historic heat wave of 2021 struck the entire Pacific Northwest, including Coeur d'Alene, into British Columbia. Extreme heat kills. According to the National Weather Service, extreme heat is the deadliest weather hazard in the U.S. (weather.gov/hazstat). Nationwide, six hundred deaths per year are reported, but the true number may be closer to 1,300 (Sarofim, M., *et al.*, 2016).

It was barely summer, late June of 2021, when temperatures in the entire region began to soar. In an attempt to stay cool, I closed the blinds and shades in my house and began running a fan early in the morning. I also stopped cooking, to keep the added heat out of my abode. I ate cold cereal, not my usual oatmeal for breakfast, then sandwiches and salads for lunch and dinner.

I began to leave early for work in Rathdrum (as a Certified Alcohol/Drug Counselor), with its air conditioning, and stayed late at night, to escape the worst of the daytime heat. However, when I returned home after 8:30 pm, it was stifling hot inside, worse than outdoors. Windows I opened found not a hint of a breeze. A fan helped a bit, but drained by the heat, I sank onto the couch instead of going for my customary walk. I headed to bed earlier than usual too, then tossed and turned, chasing elusive coolness.

I considered buying a window air conditioner. However, not wanting to worsen global warming due to the energy air conditioners use as well as "super pollutant" hydrofluorocarbons as a coolant, I convinced myself to hold off. Major heat in the past had been fairly short-lived, I reasoned as well. Not this summer, though.

After three weeks of oppressive heat, wildfire smoke returned in August. A mere haze at first, it became a pall that hung over the entire area. My eyes burned, even indoors, at work. I continued to forego my customary evening walk, further disconnecting me from the outdoors. A week later, the beauty of the hills and lake became a distant memory. My mood darkened. Feeling increasingly isolated and alone, I asked, drearily: "Will this ever end?"

Mercifully, it did. But not until we'd endured forty-four days of 90°F or hotter (nearly half the summer), between June 20th-Sept. 20th. Beginning June 24th, the heat was "unprecedented," the *Coeur d'Alene Press* reported. Ten days had highs at or above the 100°F mark. The last three days of June were the hottest on record. On the 29th the 107°F high was an all-time high for that date (Buley, 2021).

Customarily cool overnights turned uncomfortably warm — 68°F to 73°F — for five straight days, from June 28 to July 2. Temperatures of 90°F and above went on for twenty-two straight days, from June 24 to July 15. Five days were over the 100-degree mark. Average temperatures in June were over eight degrees F above normal, "the hottest late June period in recorded history," according to the *Press* (Mann, January 19, 2021).

The drought was historic as well. Rainfall from June 2nd to October 6th was just 1.3 inches, compared to a normal 5.5 inches for that time span. Only 1929, with a scant .64 inch and the big fire year of 1910 with .83 inch were drier, since 1895 (Mann, July 12, 2021).

Coeur d'Alene climatologist Cliff Harris noted the six-week period between late June and mid-August was the "hottest and driest in recorded history" in the Lake City (Mann, July 12, 2021).

Of the 129 summers on record, 2021 was the eighth hottest, comparing high temperatures for the three summer months (Mann, August 23, 2021).

While there were no reported deaths in Kootenai County from the heat wave, there were serious health impacts. Note the high number of asthma admissions at Kootenai Health from June through August 2021, in Figure 11. ACA in the heading stands for Acue Care Admissions.

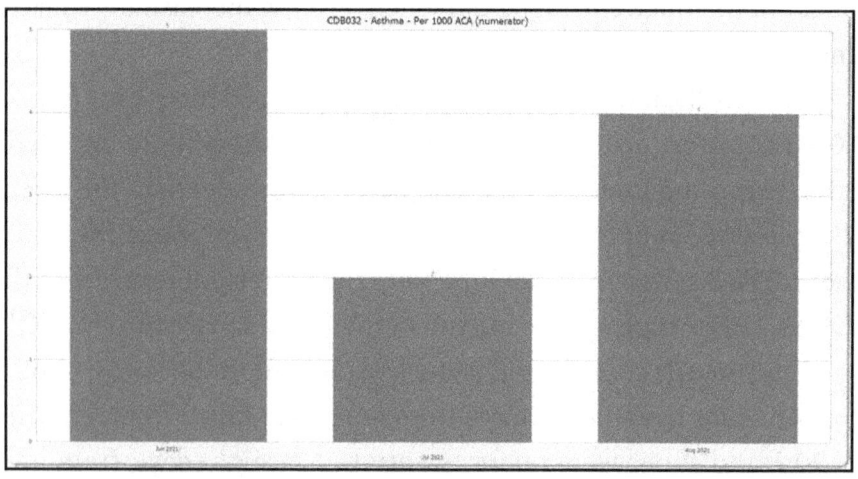

Asthma admissions at Kootenai Health, June-August 2021

Figure 11

In the midst of the heat wave, I wondered if it was being driven by climate change. A July 8th article in *The Spokesman-Review* indicated that "It ... was a 1 in 1,000-year event, according to a new analysis from an international group of researchers. The heat wave, the researchers concluded, 'was virtually impossible without human-caused climate change'"(Dreher, 2021).

The study cited in the article also offered a dire warning: "In a world with 2°C. of global warming (0.55°C. warmer than today;

[.99°F.]), a 1000-year event would be another degree hotter. *It would occur roughly every 5 to 10 years in such global warming conditions* (Philip, et al., 2021*).* [author's emphasis]

In other words, if the world doesn't keep warming below 2°C. (3.6°F) above pre-industrial levels, we could have an even hotter heat wave every five to ten years. That's a frightening and dangerous future for our children and grandchildren.

The Heat Waves of 2022

The summer of 2022 came later and was not as oppressively hot as 2021. However, there were actually five heat waves (defined by the National Weather Service as at least three consecutive days with 90°F. temperatures or higher).

Late July brought five straight days of ten degrees hotter than normal: 99°F on the 27th, 101°F on the 28th and 29th, 103°F on the 30th, then 100°F on the 31st. The second and fourth weeks of August brought more heat waves. The final one was from Aug. 30 to Sept. 2, with temperatures well above 90°F. All in all, the summer season brought twenty-nine days with highs at or above 90°F (Mann, R., 2022).

Personally, the 2021 and 2022 heat waves were a turning point. For the first time in my life, at age seventy, within months of retiring, I was concerned about my own safety. Extreme heat, after all, kills hundreds of older people like myself every year. In part to protect myself from the heat, I moved out of my trailer in October 2023.

Summer 2023 was relatively temperate in June and July. But then temperatures soared quickly in mid-August, from the mid-80s on the fourteenth to 103°F in Coeur d'Alene on the sixteenth. Six days were in the 90s. Records were broken across the Inland Northwest (Loyd & Wellford, August 17, 2023). For the entire summer, 2023 was Spokane's fifth hottest (Loyd, & Wellford, September 7, 2023).

Although air conditioning helps, it has its limits: "Standard residential air conditioners can lower the indoor temperature of a home by a maximum of 20 degrees F." Therefore, if it's 100 degrees outside, for example, A/C will lower the indoor temperature to a maximum of 80°F. Any attempt to get it cooler risks damaging the system (Apolo Heating & Air Conditioning, 2023).

We've already had days above 100°F, with more expected.

In addition, not everyone can afford air conditioning, or can afford to use it.

Later, we'll discuss an innovation that could make an important difference in people's safety.

Now, what does the future hold?

CHAPTER 7

Coeur d'Alene's Scary Climate Future

The Future Narrows

THE LEFT HALF OF Figure 3 shows the ups and downs of Coeur d'Alene's average annual temperatures, from 1893-2022. Temperatures have risen about three degrees F over that time period. Before describing our expected climate future, let's summarize the symptoms of climate change we've already experienced.

First, there's been a sharp rise in summer temperatures, as shown in Figure 10. Second, based on 2021 hospital data and the number of wildfires since 2015, there have been a high number of hospital admissions for asthma (Figure 11). In addition, less snowfall is melting earlier than in the past, and our lakes have become less solidly frozen than 100 years ago.

Toxic algae blooms are more frequent and last longer. More technical impacts, such as warmer lakes and streams, plus lower lake levels and stream flows, are beyond the scope of this book.

All of these impacts have been with a three-degree temperature increase since 1893.

Now, what about the future? Notice the sharply rising solid curve, in Figure 12. It shows the projected temperature rise from 2020 to 2100, under a high emissions scenario—the path we're on now. In 2100, the temperature would be eight degrees warmer than today: e.g., today's 72°F would become 80°F. But more threatening, the rate of increase would be **nearly three times faster** than from 1893 to 2020.

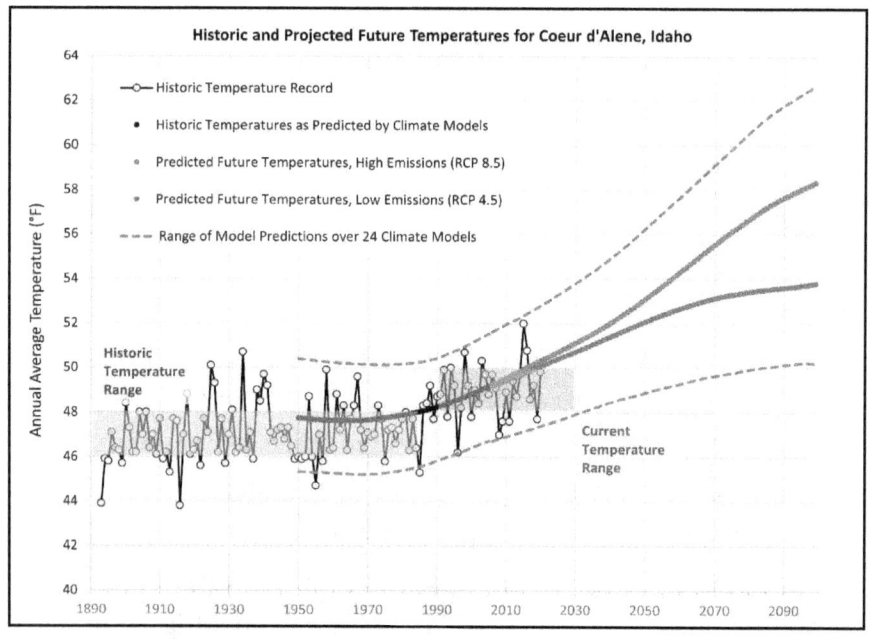

Figure 12
Source: Office of the Washington State Climatologist
https://climate.washington.edu/climate-data/trendanalysisapp/
Source: The Climate Toolbox
https://climatetoolbox.org/tool/Future-Boxplots.

What would be the consequences of such an acceleration in global warming? **What we love about living here, what makes it home for us, our town's very identity, would be eviscerated. Our home would be Coeur d'Alene in name only.**

By the year 2080, one study projected the Lake City's temperatures would become like Lewiston, Idaho today — 7.6°F warmer in the winter, and with only eleven inches of annual snowfall vs. our current sixty-seven inches. July would be 6.7°F warmer (Univ. of Maryland Center for Environmental Science). Coeur d'Alene would become, in essence, Lewiston North, its gifts of nature withered away. Our home's historically green forests, clear blue summer skies, comfortable climate and abundant, water-giving snow would be lost.

What about a more moderate warming future, shown as the blue line in Figure 12 (a low-emissions scenario)? That would require serious mitigation efforts. For instance, the increased use of non-carbon-based energy sources (e.g., solar and wind energy). Also, decreased land-use emissions (i.e., deforestation) and increased use of carbon capture and storage technology (described later). A moderate warming future would have temperatures about 4°F above those in 2020, by 2100. Remnants of our home might still exist, but barely.

To limit the very worst consequences of global warming, according to the U.N. Intergovernmental Panel on Climate Change, we need to reduce by half our country's greenhouse gas emissions by 2030. That is a tall order. However, we're making progress. Some of the promising innovations and creative solutions will be described in a later chapter.

PART II

LOCAL IMPACTS

"Not everything faced can be changed, but nothing can be changed without being faced."

James Baldwin

CHAPTER 8

Local Climate Change Impacts

"Climate change is local."
Richard Alley,
Penn St. Univ. glaciologist

AS NOTED EARLIER, ONLY local climate impacts will be discussed, those that affect us directly here in North Idaho. Therefore, important impacts such as more intense hurricanes, sea level rise, increasing ocean temperatures and acidification and loss of Arctic Sea ice, which affect us indirectly, will not be covered.

It hurts to see what we've lost, including plants and animals due to climate change, and what's at risk to be further lost. While some plants and animals may be able to adapt and survive, many won't.

Global warming has committed to extinction more than one-third of Earth's plant and animal species by the year 2050, based on current greenhouse gas emissions (Center for Biological Diversity). Loss of habitat, water resources and lack of food and water contribute to impacted food production as well.

We begin with the loss of winter ice on our local lakes,

WINTER IMPACTS

On Thin Ice

Kootenai County is home to more than twenty pristine lakes (Introduction to Kootenai County) and others less pristine. One of those, Twin Lakes, is three miles north of Rathdrum, northwest of Coeur d'Alene.

In 1972, my parents bought a cabin on the west shore of Lower Twin Lake. It was built in 1910. Shortly after its purchase, Mom met Merle Miley, the owner of the Echo Beach Resort, on the east side of the lake. He and his wife had bought the Resort in 1921.

Miley told Mom that he'd been cutting chunks of ice out of the frozen lake for decades, to sell to cabin owners and visitors during the summer (Figure 13). An underground storage shed he'd built kept the ice frozen in sawdust. Although electricity, and refrigeration, came to Twin Lakes in 1938, Mr. Miley continued to cut ice at least until the early 1970s (*In All the West No Place Like This*, p. 85). He passed away in 1979, at the age of eighty-five.

The Miley's first daughter, Loreen, was born in 1923. According to her 2017 obituary, she often told the family of her favorite Christmas, when, as a young girl, she received a new pair of ice skates. The joy of those times, skating on frozen Lower Twin Lake, is palpable in this excerpt from the obituary: "The night air was crisp, and the snowflakes gently falling down were the largest she had ever seen. She tied on her skates and ventured on the ice pushing snow in front of her blades, skating through the night with her mother" (Ellersick obituary, 1923-2017).

Lower Twin Lake encompasses only 320 acres, with a maximum depth of sixty feet. Game fish caught there include bluegill, bullhead catfish, kokanee, largemouth bass, northern pike, yellow perch, rainbow trout and white crappie. A state-record northern pike was caught in the lake in 2010 — forty lbs., two ounces in size.

Figure 13
Merle Miley (left), Howard Hensley (husband of photographer Marianne Hensley) and unidentified friend cutting ice on Twin Lakes, 1960s
(Photo courtesy of Museum of North Idaho)

Hayden Lake is 3,800 acres in size, with a maximum depth of 178 feet. About 100,000 fish per year are stocked in the lake, a smaller number to allow kokanee to grow larger. Fish stocked include bluegill, sunfish, bullhead catfish, crappie, largemouth and smallmouth bass, yellow perch, northern pike and rainbow trout.

In January 2016 the mayor of the nearby town of Hayden, Ron McIntire, had a retirement party. During his speech, the "grandfatherly gentleman" (as he was described by a Hayden staff member present at the affair) made a surprising comment to the crowd. He said he used to be able to skate on Hayden Lake growing up. But now the lake doesn't freeze over like it used to.

Lamenting the loss, McIntire conjectured out loud: "I wonder if it's because of global warming." Members of the audience were likely surprised by his musing, given McIntire's deeply conservative political and religious views. Although he quickly moved on, the retiring mayor gave voice to what many had undoubtedly noticed, even suspected.

Even smaller lakes don't freeze as often or as thoroughly as they used to. It's riskier to walk, skate or ride a snowmobile on them now, compared to decades past. As the retiring mayor recalled, Hayden Lake used to regularly harden in winter; cars and trucks drove on it regularly during the winter and ice skaters glided on its surface.

Coeur d'Alene Lake, although bigger and deeper than Hayden and Twin Lakes, used to freeze over often as well. Cars were driven across the lake as early as the 1930s, a *Coeur d'Alene Press* article noted. But it's been some time since the lake froze completely: the year 2000, it was claimed in the *Press* article (Johnson, 2019). Others say much earlier.

Coeur d'Alene Lake is the largest lake in Kootenai County, more than twenty-six miles long, with a maximum depth of 220 feet. It's surrounded by numerous parks, campgrounds, beaches, trails and recreational facilities. The southern third of the lake is owned by the Coeur d'Alene Tribe. Recommended fish to catch in Coeur d'Alene Lake include bluegill, kokanee, largemouth bass, chinook salmon, northern pike and cutthroat trout.

The thickness and extent of ice on Kootenai County lakes is an important measure of not only their long-term health but, in a broader sense, the condition of our home. Frozen lakes have defined our local winter experience: from ice fishing to ice skating, snowshoeing to snowmobiling.

Small communities in the region like Kellogg, east of Coeur d'Alene, Sandpoint, Leavenworth, Washington and others in Washington's Methow Valley depend on winter money. Winter outdoor enthusiasts fill up their gas tanks, eat and stay overnight, and use a variety of other services.

Now, however, worldwide studies show an alarming loss of lake ice. A January 2019 study, for example, showed many northern latitude lakes are at risk of experiencing ice-free winters in the coming decades. In some places, lake ice will disappear altogether by the end of the century.

Led by Professor Sapna Sharma of Canada's York University, an international team of researchers developed a model to predict which lakes worldwide would lose ice first and how that ice loss would be distributed across different latitudes. Observations of 513 lakes identified as less vulnerable to ice freezing around the Northern Hemisphere were made. A primary conclusion of the study: "an extensive loss of lake ice will occur in the next generation" (Sharma, Blasgrave & Magnuson, et al., 2020).

Although neither Coeur d'Alene Lake or other northern Idaho lakes were studied (it was, after all, a worldwide study), one image from the study showed that at 4.5°C warming above pre-industrial times (8.1°F), the lakes in our region would be totally ice-free by about the year 2100. This is consistent with the projected temperatures in the Lake City by the year 2100, under a high-emissions scenario. Temperatures would be too warm for ice to form on lake surfaces.

Such warming is considered likely if the world doesn't halve its greenhouse gas emissions by 2030, then stop them by the year

LOCAL CLIMATE CHANGE IMPACTS

2050. Fortunately, progress is being made in the U.S. and many European countries to meet this first hurdle. More on this in the solutions chapters.

A follow-up lake study, published in October 2020, found that once average winter temperatures reached 25°F, lakes were "significantly" more likely to be ice-free in the Northern Hemisphere, North America, Europe and Asia. Ice data from 122 lakes that typically freeze were studied, from 1939 to 2016.

"This isn't just happening in one lake in the Northern United States," one of the researchers stated: "It's happening in thousands of lakes around the world." Another primary conclusion: "...extreme ice-free years are becoming more frequent and severe" (Fiazolla, Blasgrave, et al., 2020). Worldwide, ice-free years are more extensive and longer lasting than in the past.

In addition to the above economic and recreation consequences, the study found the lack of ice makes the lakes more prone to toxic algal blooms. Fish and pets are harmed, even die, in those blooms. How is the lack of lake ice and harmful algae blooms connected?

Professor Sharma said lake ice serves as an ecological "reset." Lakes are warmer in years without ice cover, and they stratify earlier, which could make them more prone to toxic algal blooms (Sharma, et al., 2019). Fish die in those blooms, when dissolved oxygen levels drop too low. Also, warm water in the summer holds less dissolved oxygen than cold water (U.S. Geological Survey, Water Science School, 2019).

Figure 14 shows winter air temperatures for nine Kootenai County lakes, from 1971-2000, then projected to the year 2100. By the 2010-2039 time period, winter temperatures for all nine lakes are projected to rise by three degrees F over 1971-2000 levels. The first temperatures shown are based on a high-emissions scenario; the second, a low-emissions scenario.

This information references a key point from the 2020 study: when winter (Dec.-Feb.) air temperatures are above 25°F. Northern Hemisphere lakes are "significantly" more likely to become ice-free. If we apply that information to the nine Kootenai County lakes in Figure 14, only Lower Twin Lake was below 25°F. in 1971-2000, more than twenty years ago. Most lakes were significantly higher in temperature than 25°F.

	1971-2000	2010-2039	2040-2069	2070-2099
Avondale	30.2	33.2/32.6	36.1/34.9	39.7/35.9
Coeur d'Alene	31.1	34.2/33.5	37.0/35.8	40.6/36.7
Fernan	30.2	33.3/32.7	36.1/34.9	39.7/35.8
Hauser	29.9	33.0/32.3	35.8/34.6	39.4/35.5
Hayden	30.2	33.2/32.6	36.1/34.9	39.7/35.9
Lower Twin	**22.6**	25.5/25.0	28.4/27.1	32.1/28.2
Pend Oreille	29.4	32.5/31.8	35.3/34.1	38.9/35.0
Priest	27.2	30.2/29.6	33.1/31.9	36.8/32.8
Spirit	28.3	31.4/30.7	34.2/33.0	37.8/32.7

Figure 14
Source: Northwest Climate Toolbox (climatetoolbox.org)[1]

The key takeaway: all nine Kootenai County lakes are "significantly" more likely to become ice-free, at least by the year 2100, perhaps sooner. **Human-driven climate change is causing us to lose our frozen winter lakes.** Our great-grandchildren are likely to have no place to ice skate outdoors, ride snowmobiles or ice fish as adults. Even under a lower emissions scenario, with substantial reductions in winter temperatures, they are expected to remain far

---

[1] A collection of online tools to visualize past and projected climate and hydrology of the Northwest.

above the 25°F. cutoff to maintain lake ice. Only the small Lower Twin Lake might be spared.

When would our lakes permanently lose their ice cover? No one knows. It depends on our present and future level of greenhouse gas emissions. What we do know, though, is that our local lakes are on a path to being ice-free, something that has never happened in recorded history. It would be a tragic loss for our home and our cultural heritage.

Even more, it would likely result in a dramatic increase in toxic algae blooms, seriously endangering fishing, swimming and the stability of the entire lake ecology. What we still have control over, though, is how quickly that happens, or even, long-term, possibly reversing the trend.

Ice On Lakes Is Not Just for Having Fun

I spent many joyous days growing up swimming in Coeur d'Alene Lake and Lower Twin Lake with my brother and sister. Later, as an adult, I swam and water skied on Lower Twin. Dad enjoyed grabbing logs from the lake in the early spring to dry on shore, for firewood later. He also appreciated working on projects at the cabin for a week or two by himself.

In later years, my brother and I water skied on, and swam in, Lower Twin. Later still, I did the same with my two sons. Two weekends a summer, for years, sister Janet shared the peace and quiet with friends from Seattle as well. Her son, Dylan, loves to fish there, and has made notable upgrades to the cabin's livability. My oldest son, Ryan, spends weekends in the summer and a few days in the winter with his dog and friends at the cabin, joined occasionally by his brother Justin.

I spent a fun weekend at the cabin with my brother and sister during one of the hottest periods of the 2021 summer. Sitting on the early morning, sun-drenched porch, watching a muskrat carry sticks to its nest and a gaggle of turkeys scurry to the lake for an early morning drink, feeds the soul.

Conditions in the lake, however, have changed. There's been a sharp rise in the number of water lilies and other plants growing in it. The overgrowth is most noticeable in the canal in front of our boat shed, as well as the channel between the Upper and Lower Twin Lakes and near our cabin dock. Boat propellers can get clogged with the plants and swimming is no fun among them.

An 11.5 mph current carrying nitrogen and phosphorus from Upper Twin Lake is what's causing the enhanced plant growth (2022 Twin Lakes Improvement Association meeting), along with increased lake temperatures. Dredging the Lower Twin Lake channel during the fall of 2022 helped to reduce dense plant growth. The same is proposed for the canal in the fall of 2024.

As winter ice is lost earlier in the spring on most lakes, the waters are exposed to the sun's warming rays for longer periods of time. Evaporation increases, lowering lake levels. Water temperatures then stratify lakes into different temperature layers (warmest on top; coldest on the bottom) earlier and longer than before. This reduces the availability of dissolved oxygen and nutrients for fish and other aquatic life.

At its extreme, anoxia, the loss of oxygen, can result, suffocating fish. Anoxia occurred earlier, and lasted longer than ever before in Coeur d'Alene Lake during the summer of 2015. Fish deaths increased as a result.

A twenty-one-year-old I interviewed in June 2022, Ryan Hanna, offered a personal story of warmer lake waters. As a seasonal invasive species surveyor with the Dept. of Agriculture, he said he felt the "absolutely sweltering" 115°F temperatures on the Dworshak

Reservoir (on the North Fork of the Clearwater River, south of Coeur d'Alene). During 2021's "hyper intense" heat wave, driven by climate change, he had to cool off every fifteen minutes. Since its first six feet were still at 80°F, he had to dive deeply into the lake to do so (Hanna 2022 interview).

Warmer lakes, it could be said, are sicker lakes.

Harmful Algae Blooms

Warmer waters, along with the nutrients from fertilizer use, promote the growth of harmful algae blooms. These blooms occur in warm, slow-moving waters, along shorelines in the summer and fall. In sufficient concentrations, the blooms produce different colors of cyanobacteria, which produce a toxin that can be deadly to animals and make humans quite sick. Pets, children, the elderly and people with compromised immune systems are the most at risk of harmful exposure (DEQ website).

When a bloom is confirmed to be harmful by the Idaho Department of Environmental Quality, a public health advisory is issued. The advisory urges the public to be cautious recreating in or near a body of water with slimy, oil-colored scum on the surface, especially when ingestion is a risk. Keep dogs completely away from these areas.

The first recorded toxic algae bloom in northern Idaho was in 1985-1986, in Black Lake, a small chain lake east of Coeur d'Alene (DEQ information). They have since become a yearly occurrence. Other small North Idaho lakes, including Avondale, Cocolalla and Spirit have had health advisories issued due to algae blooms.

Fernan Lake is one of the most highly impacted water bodies in the state. In 2015 residents and lake managers asked Frank Wilhelm, with the Department of Fish and Wildlife Sciences at the University of Idaho, to investigate the cause of the blooms. Water and nutrient flows into and through Fernan Lake were studied as part of a National Science Foundation project. They found that eighty-one percent of the phosphorus — a nutrient that stimulates algae growth — entering the lake never leaves, allowing the phosphorus to build up year after year (LaCroix, 2015).

The June 2022 interview with Ryan Hanna was informative about Fernan Lake's recurring toxic algae blooms. A recent University of Idaho graduate in Environmental Science, Hanna wrote his senior thesis on HABs in North Idaho.

Blooms in Fernan Lake are primarily a land use, not a climate change problem, he said. Why is that, I asked. Since Fernan Creek, which feeds the lake, is no longer meandering as it used to. In the past there had been plenty of vegetation on its banks to hold back phosphorus-rich fertilizers from higher elevation agricultural lands. Now the creek is completely channeled, so all of the fertilizers go straight into the lake. Plus, with warmer summer waters, cyanobacteria blooms peak, between 70-75°F, and produce toxins.

Hanna explained that the lake's east-west orientation has the wind constantly mix its layers, causing it to be ten to fifteen degrees warmer than Coeur d'Alene Lake. Warmer conditions allow invasive species to thrive there as well.

What's the solution, I asked? Replant vegetation and re-meander the stream, to catch the phosphorus, he said. Unfortunately, later in 2022, Fernan Village declined the Department of Ecology's plan to help ameliorate the problem.

Public health advisories due to toxic algae blooms are now common. In 2019, advisories were issued for Hayden and Fernan Lakes, as well as Lower Twin Lake, not near our family cabin. Although only three were issued in 2020, one for Fernan Lake, from July 20th

to Nov. 12th, lasted for nearly four months (2022 telephone conversation with DEQ surface water manager Robert Steed).

The summer of 2021 had an historic number of toxic algae blooms. Eleven water bodies were issued health advisories due to Harmful Algal Blooms (HABs), compared to just three in 2020. In the words of Chantilly Higbee, Department of Environmental Quality's Water Quality Compliance Officer in Coeur d'Alene: "This year [2021] was unprecedented for the number of water bodies we issued advisories for" (Higbee, 2021 email).

Algae blooms present problems even when they die. When they settle to the bottom they are devoured by microbes, which further deplete oxygen levels. Lake health further deteriorates.

Nearby property values plummet as well. Losses near lake homes were between eleven percent and seventeen percent, rising to above twenty-two percent for lake adjacent homes across six Ohio counties (Wolf & Allen Klaider, 2017). Anecdotal evidence "across multiple states" even suggest a thirty to fifty percent drop due to the bad smells and toxins emanating from HABs (Arenschield, 2015).

That brings up one final lake impact: exacerbating the spread of non-native species. These include European Watermilfoil and quagga mussels, which clog drains, drinking water systems, irrigation pipes and dams. Research indicates the situation is only likely to worsen, with future changes in temperature and precipitation (Gervais & Kovich, et al., 2020). These invasive species can quickly overtake native species, seriously disrupting a lake's ecosystem.

Fortunately, as of our June 2022 interview, there were no Zebra mussels in Idaho, Oregon, Washington or Montana waters. This is due to the effectiveness of water craft inspections at state border crossings, Hanna said.

THE DEATH OF WINTER?

Less and Earlier Melting Snow

Figure 15 shows no clear trend in snowfall amounts in Spokane (which has more reliable snowfall records than Coeur d'Alene), from 1893-2023. The use of high-speed computers in a 2020 study, however, indicated "widespread reductions in snowfall are evident across Idaho, with *reductions of up to 15% in snowfall in the Bitterroot Mountains* during the period 1950-2020 (Lynn, et al., 2020). [author's emphasis] The Bitterroots are east of Coeur d'Alene, along the Idaho-Montana border.

The future is expected to see even more pronounced loss of snowfall in the Coeur d'Alene area. With projected temperature increases, the snow drought during the winter of 2014-2015 is expected to become more frequent and multi-year in nature.

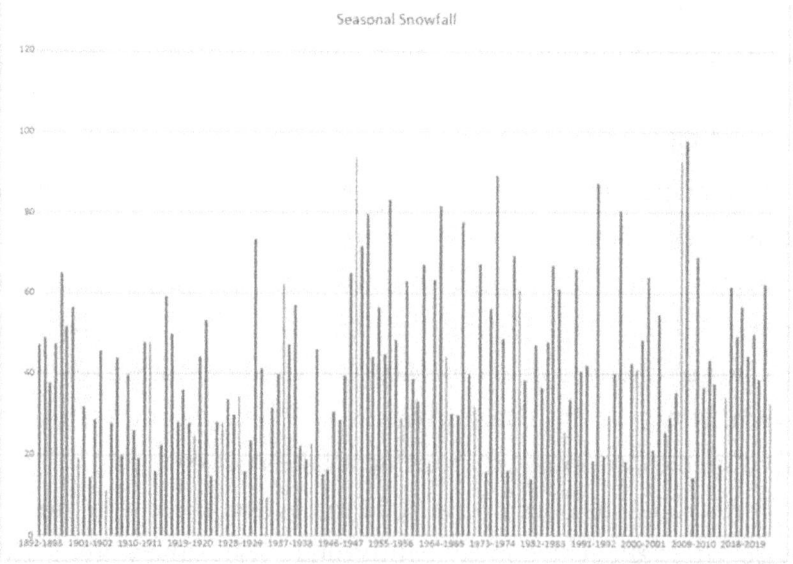

Figure 15
Seasonal snowfall totals, 1893-2023; Spokane, WA data
(Source: National Weather Service -- Spokane, WA office)

LOCAL CLIMATE CHANGE IMPACTS

Similar to the very low snow amounts in 2015 which caused two ski resorts to close early ...

> multiple consecutive years of snow drought are projected to become much more common, particularly in the northern part of the state. Across parts of Idaho that averaged 4 inches or more of annual maximum SWE [Snow Water Equivalent, the amount of water in snow] during the late 20th century [e.g., in the Coeur d'Alene area], multi-year snow droughts occurred in approximately 7% of those years; multi—year snow droughts are expected to occur 45% of the time by the period 2050-2079 in a high-warming scenario (Marshall, et al., 2019).

In addition to declining snowpack, the snow is melting sooner, due to warming spring temperatures. How much sooner?

On average, a 2010 study showed snow melt was one to two weeks earlier, compared to the 1975-2010 time frame (Marshall, et al., 2019).

What's the impact of earlier snowfall runoff? Higher-than-normal flows in the winter months and lower-than-normal flows in the summer months when it's most needed for irrigation, power generation, and recreation. Also, soils and vegetation dry out sooner, weakening trees to fight off tree-killing pests. The result: even more dead vegetation that burns easier, increasing wildfire danger.

In short, our less healthy lakes are being fed by early snow melt, with less late summer water available.

Will Winter Skiing Survive?

This situation has important implications for skiers, snow boarders, snowmobilers and snowshoers, as well as the communities dependent on their income. Snow sports have been a part of our home since the early 1900s, and a factor in home values and jobs.

Nationwide, the snow sport industry has been affected by less, and earlier melting snow. Due to increasing temperatures, the ski season in the U.S. is opening nearly a month later than the 1960s or '70s (McCusker & Hess, 2018).

A 2017 study of the projected impacts of climate change on skiing and snowmobiling in the U.S. had some chilling news. Two hundred and forty-seven U.S. ski resorts, including Lookout Pass, Schweitzer Mountain and 49 Degrees North in our region were studied. The number of days in the winter ski season in 2050 and 2090 were compared to the baseline of 2011.

Below is a summary of the projected drop in the length of future ski seasons for the three resorts.

Percent drop in number of ski season days (compared to 2011)

	2050		2090	
	Emissions Scenarios			
	Low	High	Low	High
Lookout Pass	29	68	68	95
Schweitzer Mountain	46	62	73	96
49 Degrees North	23	40	42	85

(Source: Wobus, Hosterman, et al., 2017)

What do these findings tell us? The way it's going, unless we cut greenhouse gas emissions in half by 2030, then to net-zero by 2050,

LOCAL CLIMATE CHANGE IMPACTS

there will be **no ski seasons at either Lookout Pass or Schweitzer Mountain by 2090 and a severely limited one at 49 Degrees North**. This is based on a 95%, 96% and 85% drop in length of ski seasons under a high emissions scenario.

Lookout Pass and Schweitzer Mountain would have only **five or six days** to their ski season by 2090; 25 days at 49 Degrees North, hardly economically viable. The low emissions scenario, on the other hand, would likely keep each resort functioning, at least partly (Wobus, Hosterman, et al., 2017).

While not in the study, Silver Mountain Ski Resort *might* fare a bit better. However, that estimate is based on a less precise length of ski season (number of days between the first fall freeze and the last spring freeze) and a different baseline time frame (1971-2000) than the national report. This estimate projects Silver Mountain would lose 1.5 months of its ski season by mid-century and more than two months by 2090, under a high emissions scenario.

If, however, those emissions dropped to the low scenario, the losses would be significantly less — four days by mid-century, three weeks by 2090 (based on data from the Northwest Climate Toolbox).

Schweitzer Mountain has its own climate-limited future. Its early January snowpack has decreased at a rate of over 1.5 Snow Water Equivalent [amount of water in snow]-inches per decade (McCusker & Hess, 2018, Figure 3). To understand what that means, an old rule of thumb was that for every ten inches of snow there would be one inch of water (National Weather Service). However, with so many variables involved now, including temperature, depth and density of the snow, it applies in only a very general manner.

Outdoor winter sports, long enjoyed by many here, and a great asset to our economy, will have a perilous future in a climate-impacted Inland Northwest.

North Idaho Wildfires

> "...human-caused climate change caused over half of the documented increases in fuel aridity since the 1970s and doubled the cumulative forest fire area since 1984."
>
> (Abotzoglou & Williams, 2016)

Higher temperatures dry out vegetation and forests, lengthen the wildfire season and significantly increase the number of critical wildfire danger days. For every one-degree rise in temperature, fire risk is increased by up to twenty-five percent (Wuerthner, 2022).

A friend's story will describe the recent effect of wildfires and smoke on our home.

David Muise is not the first, and won't be the last Californian to move here to get away from wildfires in the Golden State. Dave and his wife moved to Coeur d'Alene from the Santa Rosa area in 2017, in part due to the Tubbs Fire, at the time the most destructive in California history. Twenty-two people were killed and over one billion dollars in economic losses were sustained (Wikipedia).

For one week, Dave said, they had their car loaded, waiting to be told when to evacuate. Although the call never came, they left anyway, due to both "crazy" home prices and the danger of perpetual wildfires. That danger got very real when the burned black leaves from bay trees, miles away, became embers near their residence.

There's a tragic irony to this and similar stories. Nearly every summer since they moved here to get away from the wildfires and smoke, they've had to endure them here.

The summer of 2021 was the worst. During the week his daughter and family friends visited there was only one "clear" day, he said. Every other day they had to spend indoors. An outdoor paradise became an indoor prison. As a result, the family friends said they'd never return. Likely, their story is not unique.

For what turned out to be nearly two weeks during that summer, and five of the previous six, our home was taken from us. Human-driven climate change was the thief.

U.S. Forest Service data shows that wildfires are getting bigger and lasting longer: the number of large fires (more than 1,000 acres) has tripled since 1970 and the duration of the fire season has grown by more than 100 days. There are **more fires** as well (sixty-one percent of fires in the Western U.S. have occurred since 2000) and they are **larger** (since 1950, the number of acres burned per year has increased by 600 percent) (Abotzoglou & Williams, A.P., 2016).

Plus, the National Oceanic and Atmospheric Administration (NOAA) predicts by mid-century (2041-2070), the conditions for "very large fires" will substantially increase throughout the Western U.S. (NOAA, global science). Figure 16 shows eastern Washington and northern Idaho will have a 300 percent to 400 percent increase in these very large fires.

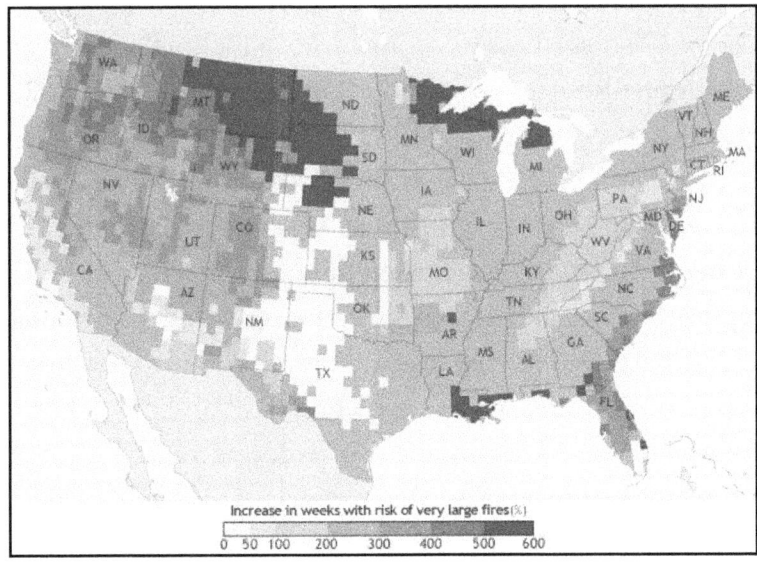

Figure 16

During the summer of 2021, four fires over 1,000 acres struck the Idaho Panhandle National Forest (IPNF), which includes forests in the Coeur d'Alene area. The fires were reported to be caused by lightning storms that had passed over the IPNF in early July ("Fires prompt temporary closures," Idaho Panhandle Natural Forest, 2021).

By the end of the month, Forest Service Supervisor Carl Petrick noted: "Hot, dry and breezy conditions have contributed to increased fire activity.... This level of fire activity is not normally seen in our area until August. Recent record-breaking temperatures and an exceptionally dry winter and spring are causing extreme fire conditions."

Petrick predicted a "lengthy fire season ... with "the potential to last until November" ("Fires prompt temporary closures," 2021). Luckily, with cooling temperatures in August, by September the threat of continuing wildfires had abated. I was convinced enough of the danger of wildfires that I got up on the roof of my house to cut down overhanging tree limbs.

Figure 17 shows heavy wildfire smoke looking south from the Coeur d'Alene Resort Boardwalk. It was taken on August 2, 2021 by David Muise.

The first half of the 2022 wildfire season here was uneventful due to cooler and wetter conditions. Warmer and drier conditions didn't begin until late June, then again in late July, ramping up through much of August. No threatening wildfires, though. We were fortunate. Wildfire smoke didn't return until the first week and a half of September. It wasn't significant and didn't last long.

August of 2023 was an entirely different matter — not only uncomfortably close wildfires but hazardous air quality from those blazes and Canadian wildfires. First, the 3,000-acre plus Ridge Creek fire northeast of Hayden Lake was still burning at the end of the month. Then, on August 18th, the Gray Fire forced the evacuation of Medical Lake, Washington, forty-five miles west of the Lake City. Further north, the Oregon Road fire in Elk, Washington erupted as well. The pair of fires took two lives and destroyed hundreds of homes.

On the nineteenth, Spokane had the worst air quality in the nation (Lagoo, KREM-TV, 2023), with an unbelievably high AQI of 448 — hazardous to anyone's health to be outside. The AQI at the Coeur d'Alene Costco was such a concern at noon when I went to get groceries, it was written on a white board outside: a very unhealthy 280. Blankets of eye-stinging smoke covered the area, similar to Figure 17's photo.

Figure 17
(Photo courtesy of David Muise)

Health Impacts of Climate Change

"A stable climate is the most fundamental determinant of human health."
(Mailbach, E., Miller, J., Armstrong, F., et al., 2020)
2020 Health Professionals article

When I first showed the movie, "An Inconvenient Truth," about the threat of global climate change to my parents, Mom said: "I'm glad I'm old so I won't be around to see that stuff happen. That's scary." Fortunately for Mom, an RN, she passed in 2013, two years before wildfire smoke, and its health impacts, brought climate change too close to home.

Mom and Dad retired to Coeur d'Alene in 1987. Soon after, Mom began to help develop what became the Dirne Clinic (now Heritage Health), providing health care to low-income patients.

Wildfire Smoke

Locally, smoke from wildfires is the most noticeable sign of climate change. When the eye-stinging gray haze first appeared in the summer of 2015, it surprised me: "What's this? This isn't supposed to be happening here." Field burning on the Rathdrum Prairie, after all, had stopped ten years earlier. When it persisted beyond a couple of days, I grew worried: "What's happening here?"

No smoke the following summer raised the hope that it had been a fluke of nature. As we all know, though, that was not the case. Only one other summer (2022) of our past nine had little or no smoke. In 2023 it seemed as if every place in the country was on fire.

Wildfire smoke isn't just ugly, it's hazardous, containing hundreds of chemical compounds. The smallest particles in smoke (fine particulate matter, one-thirtieth the width of hair) become embedded deeply into the lungs and bloodstream. By far the biggest source of those tiny particles in the Northwest is wildfire smoke. Since 2011, thirty-four of Kootenai County's thirty-five unhealthy air quality days occurred during the wildfire season (June-September) (Table 2, EPA Air Data, 2021).

From 2019 to 2021, an American Lung Association report found that the Coeur d'Alene-Spokane Valley-Spokane area had the fourteenth most polluted air in the country. That's almost entirely due to fine particulate matter in wildfire smoke. Yakima, WA was number 11 and Missoula, MT number 15. Bakersfield, CA had the most polluted air (Leber, 2023).

Scientists link smoke exposure with long-term health problems, including decreased lung function, heart conditions, weakened immune systems and higher rates of asthma. The pall of smoke also aggravates Chronic Obstructive Pulmonary Disorder (COPD) and heart failure conditions, even in people hundreds of miles from the flames.

Some have to flee the area when smoke invades. High-risk groups, including children and the elderly, are most affected. They may seek relief from their rescue inhalers, visit their pulmonologist, cardiologist or allergist more often or have to be rushed to the Emergency Room.

With high smoke levels, everyone can be isolated indoors for days or even weeks, away from the smoke and any outdoor activity. Some become depressed and/or anxious, feeling trapped. Those with pre-existing mental health conditions worsen. Sleep problems are aggravated. Even those who are perfectly healthy can develop serious health problems due to extended wildfire smoke exposure.

The impacts of breathing wildfire smoke in the Inland Northwest have been noted in the past few years. In the summer of 2018, for example, Spokane internal medicine specialist, Dr. Matt Hollon, "saw a lung disease patient develop an acute worsening of respiratory symptoms and an otherwise stable patient suffer a heart attack one day after severe smoky air conditions." Even back then, the doctor said he "sees a climate change correlation" (Lind, 2019).

How can air pollution and tiny particulate matter cause strokes and heart attacks? Dr. Hollon explains: "Tiny particles get in lungs and cause inflammation in the body, ... and that can become a trigger for other inflammatory reactions in the body contributing to those cardiovascular incidents" (Lind, 2019).

In June 2019, several medical groups, including the American Medical Association and American Academy of Pediatrics, issued a call asking the government and other leaders to recognize **climate change as a health emergency** (Lind, 2019).

Closer to home, the town of Seeley Lake, near Missoula in western Montana, was inundated with smoke for seven weeks in 2017. A three-year case study of the effects of wildfire smoke on human health in Seely Lake was completed in 2020. Doctors were surprised how long the consequences of the haze lasted.

The study showed "people's lung capacity declined in the first two years after the smoke cleared. Chris Migliaccio, an immunologist with the University of Montana and his team found the percentage of residents whose lung function sank below normal thresholds more than doubled in the first year after the fire and remained low a year after that" (Houghton, 2020).

A recent study attributes 20,000 premature deaths in the United States in 2018 to those small smoke particles in the air from human-caused fires (Carter, 2023).

Wildfire smoke as a health issue in the Inland Northwest is quite recent. According to an article in *The Spokesman-Review*, the Regional Clean Air Agency began monitoring tiny air particles (called $PM_{2.5}$) in Spokane in 1999. That year there were no days when particulate matter exceeded the threshold for health concerns. Every single day the air was safe to breathe.

In the fifteen years between 1999 and 2014, there were thirty-three days of unhealthy air quality for vulnerable people. Only three of those were attributed to wildfire smoke; the rest were to wood-burning stoves in the winter, emitting smoke that became trapped when the atmosphere was stagnant.

The year 2015 began our string of bad air quality days. Not only was it Washington state's worst wildfire season in history (until 2020 broke the record), it was also the nation's, with more than ten million acres burned around the country. Plumes of smoke, carried hundreds of miles by the wind, settled in our lungs. Coughing, choking and stinging eyes ensued.

In 2015, Spokane County's air quality exceeded the safe limit sixteen times. On eleven days the air quality was "unhealthy for some." Five days were unhealthy for everyone and for the first time, unhealthy days happened in August.

There were no unhealthy air quality days in 2016. However, during eighteen days in 2017 the air was not healthy for some to breathe, including ten days in August and a five-day period in September. That included the worst stretch of air quality in twenty years.

In the Lake City, students had to be kept indoors the first few days of school that September, according to Coeur d'Alene School District Director of Communications Scott Maben. There were no outdoor recesses or P.E. and all athletic practices were held indoors (June 2022 telephone communication with Maben). A Lake City High School football player at the time, Ryan Hanna, said "a couple" of football games were canceled as well (June 2022 Ryan interview).

On Sept. 5, 2017, air pollution reached the highest level, the "very unhealthy" category, for the first time in Spokane since 1999 (when there was a dust storm on Sept. 26th). There were fourteen days of unhealthy air that year.

In 2018, there were thirteen days above the moderate air quality level. Only two days were in the unhealthy range. However, from August fifth to August seventh, Spokane had the worst air quality in the nation, according to a report on KISC-FM radio.

To summarize: thirty-three unhealthy air quality days in sixteen years, from 1999-2014. That's about two a year. Then, forty-eight unhealthy air quality days in only five years, from 2014-2018 (Vestal, 2019). That's nearly ten a year.

According to the Spokane Regional Clean Air Agency website, there were two total days of unhealthy air quality for sensitive groups in 2019. In 2020 there were five, with one at a very unhealthy level. The summer of 2021 included the historic heat wave and air quality was the unhealthiest since 2018. There were eight days total, two for sensitive groups, six at the unhealthy level (Spokane Regional Clean Air Agency website). That's eight unhealthy air quality days in one year.

On Sept. 12, 2021 multiple fires in the region caused the Air Quality Index to hover between 160 and 180 in north Spokane. In the evening it reached 213, in the "very unhealthy" category. In Coeur d'Alene, the school district kept students indoors for recess, P.E. and athletics (June 2022 phone conversation with Maben).

Many undoubtedly worried: ***What's happening to our home?***

Asthma

Figures 18 and 19 show yearly hospital admissions for asthma and COPD, respectively, at Kootenai Health hospital in Coeur d'Alene, from 2015 through June of 2021, half the year (July 2020 email from Kelly Fry, Kootenai Health Media Dept.). ACA in the heading stands for Acute Care Admissions.

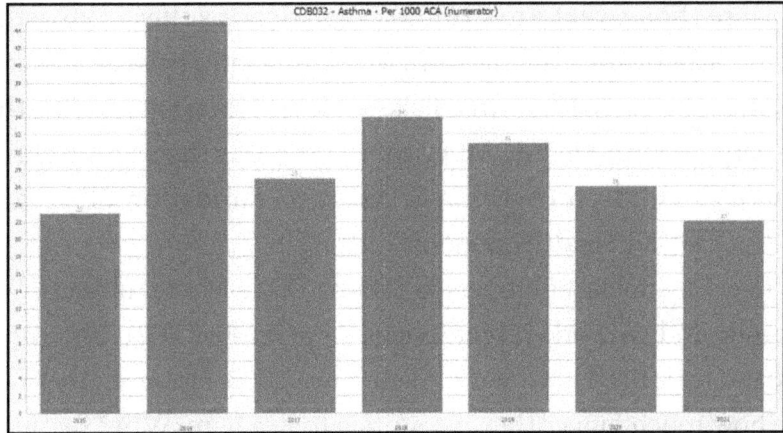

Figure 18

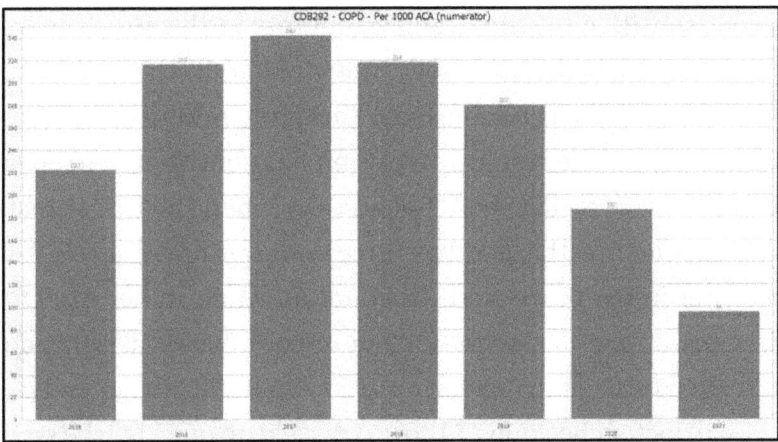

Figure 19

Asthma is one of the most common chronic diseases in our country. It costs eighty-two billion dollars a year in medical costs, loss of work and missed school days (Asthma & Allergy Foundation of America, 2018). Asthma causes swelling and narrowing of the airways, leading to trouble breathing, sneezing, coughing and tightness in the chest.

More than twenty-five million Americans suffer from asthma and ten Americans a day die from it. Asthma rates have been increasing since the early 1980s in all ages, sex and racial groups (Asthma & Allergy Foundation of America, 2018) and is the leading chronic disease in children (Asthma & Allergy Foundation of America, 2018, Asthma | Healthy Schools).

These are not just numbers to Tabitha Day, a Master of Social Work therapist and co-worker, I interviewed in October 2021. She and her husband have lived in Coeur d'Alene since 2003. However, they're not sure how much longer they'll continue to live here. Why? Because of her asthma, aggravated by persistent wildfire smoke the past few years.

Day reports struggling to breathe if there is any wildfire smoke: "I'm very sensitive to smoke and pollutants in the air." If there are any wildfires or if there's "super heavy pollen in the air, I'm really stuck at home." She is unable to go camping, hiking or swimming, which she loves. The resulting isolation has contributed to episodes of depression, as well as bouts of anxiety when she can't breathe.

To help her breathe, Day has three home air filters, a HEPA filter, an air filter in her office, and the conditioned air in her car. During wildfire season she's had to pay for four rescue inhalers, at eighty dollars apiece, plus the need to regularly use an Albuterol inhaler and pay for additional doctor check-ups. "Those costs add up quickly," she stressed.

Many of us have become familiar with the Air Quality Index (AQI), which measures the level of air pollution and commensurate health concerns. A level above 100 micrograms per cubic meter (on a 0 to 500 scale) causes Day to struggle to breathe, while the general public is less likely to be affected. Anything over 150 is considered unhealthy for everyone.

At a level of 151-200 she can't go outside except for brief moments. Some members of the public may also experience health effects, even without underlying health conditions. Levels approaching 300 have occurred here in the past three years, Day said, which would be an emergency health warning for everyone.

Being a smoker for twenty-five years, and unknowingly exposed to airborne black mold where she previously lived, contributed to warnings from her doctor about early onset COPD or emphysema. Since she stopped smoking, Day said she's had less lung infections and has been able to hike outside with her husband and dogs on warm, good air quality days.

Asked if she's seen any changes in the climate since she moved here, she quickly answered: "Oh yeah, a significant change." In the past we had "normal spring and summer weather." However, in the past five years "it's gotten hotter and hotter," and her "perception" is we're having "less severe snowfall" and "not as much rain" in the spring. In addition, the speed with which the climate is changing is "way faster than it should be."

Finally, heart failure patients can die prematurely from wildfire smoke. Nationally, increases in particulate matter concentrations in the air (e.g., from wildfire smoke) are a "significant mortality risk factor for heart failure patients," according to a twelve-year study among over 20,000 heart failure patients. The study added: "Elevated $PM_{2.5}$ exposures result in substantial years of life lost even at concentrations below current national standards" (Ward-Caviness, Yasdi, Moyer, et al., 2021).

Local health impacts can also be measured by the number of visits to Kootenai Health's Emergency Department (ED) and Marimn Health, the Coeur d'Alene Tribe's community health center.

Figure 20 shows the number of visits to Marimn Health for asthma, COPD and heat-related conditions (heat stroke, heat exhaustion and dehydration), from 2015 to 2023 (Laumatia email, April 3, 2024).

Dx	2015	2016	2017	2018	2019	2020	2021	2022	2023
Asthma	12	30	95	105	77	97	99	90	96
COPD	35	111	131	134	163	173	161	183	193
Heat dx	4	13	35	25	17	16	14	18	33

Figure 20

Note the dramatic increase in the number of visits for each condition since 2015. All but 2016 and 2022 were years with wildfire smoke while seven of the last nine summers have been unusually hot.

Data on air quality-related respiratory illnesses, asthma and heat-related visits to Kootenai Health's Emergency Department, from 2019 to March 2024, are summarized next.

Air Quality-Related Respiratory Illnesses

Wildfire smoke causes, or exacerbates, most respiratory illnesses, including COPD, bronchial asthma and emphysema. From January to September 2019, respiratory illnesses comprised five to six percent of all ED visits at Kootenai Health. That number jumped to fifteen to twenty percent through the COVID-19 pandemic, beginning in March 2020. Rates stayed high until diagnosis codes were changed in February 2022, to provide greater specificity between asthma, influenza and COVID-19 symptoms.

Asthma Emergency Department Visits

Asthma ED visits jumped from one to two percent of all ED visits in January 2019 to four to seven percent in September 2019. Visitation rates stayed high until respiratory codes were changed in February 2022, as noted above.

Heat-Related Illnesses

In 2019, sixteen Emergency Department visits were for heat-related illnesses; 2020 had twenty-eight visits. The heat wave-related illnesses of 2021 were off-the-charts, with seventy-two visits, more than twice as many as any other year. The years 2022 and 2023 had twenty-three and eighteen ED visits, respectively (Rich, April 3, 2024 email). Heat-related illnesses include heat stroke, heat exhaustion, heat cramps, heat rash and hyperthermia (abnormally high body temperature due to a failure of the body's cooling mechanism).

This information documents the impact of an unstable climate on our physical health. Too many of us have to seek relief from conditions we're not used to experiencing, in a home that until recently had nurtured and supported us. Changes in our home haven't hurt us in the past, but increasingly do so now.

Allergies

Wildfire smoke is not the only air quality issue in the Coeur d'Alene area. Common pollen from ragweed, trees and grasses effects many people's health as well. Rates of allergies and hay fever have increased in the Coeur d'Alene area, in part due to the disrupted climate. Warmer temperatures spur plant growth, which spew more pollen into the air.

Allergies cause eyes to become irritated and puffy, noses stuffed and runny, throats inflamed and itchy. We sneeze and sniffle a lot. My former wife Lisa's allergies were so bad in 2016 she could barely walk up the stairs in our split-level home. She felt so weak she was afraid she was dying. Thankfully, she's better now, but April and May still remain the worst months.

Allergies cause people to miss work and reduce productivity. Children and adults miss school and work, which affect performance.

Treating the symptoms raises medical costs, including emergency room and doctor's visits as well as the use of inhalers or other medications. Anxiety flares, understandably, when breathing is restricted during an allergy attack.

Research has found pollen seasons to be twenty days longer, with a twenty-one percent increase in pollen concentrations across North America between 1990 and 2018 "...strongly coupled to observed warming. Human forcing of the climate system contributed ~50% of the trend in pollen seasons and ~8% of the trend in pollen concentrations."

Higher pollen counts, due substantially to climate change, can increase pollen sensitization in children, leading to an increase in adolescents and adults with allergic asthma (Anderegg, Abatzoglou, Anderegg, 2021, 5). Which plants saw the biggest increases in growth and pollen production from temperature rise? Ragweed and tree pollen (Anderegg, Abatzoglou, Anderegg, 2021).

Pollen counts here at home can be high in both the spring (March-May) and fall (Sept.-Nov.). To see if the previously noted study results correspond to Coeur d'Alene's experience, I looked at temperature changes in Coeur d'Alene, from 1969 to 2018. Would the rise in those temperatures result in increased allergy and/or asthma symptoms? I wanted to know.

July and August have shown big temperature jumps, which continued into September, October and November. Surprisingly, January had the single largest temperature increase; spring temperatures the lowest.

Next, would local allergy and asthma cases correspond to these temperature increases? Unfortunately, I ran into resistance from the medical community and various recreational and hospitality companies in trying to obtain local perspectives. Kootenai Health, however, provided five years of information, seen in Figures 18 and 19.

How do major cities in our region rank nationally, in asthma and allergy rates, illness severity and availability of treatment services? All are considered "average" by the Asthma & Allergy Foundation of America. Spokane is number thirty-eight of the top 100 Asthma Capitals; Boise is number fifty-five and Seattle number sixty-four (Asthma & Allergy Foundation of America. Asthma Capitals). Of the top 100 Allergy Capitals, Spokane is number eighty (Asthma & Allergy Foundation of America. Allergy Capitals).

Higher temperatures and wildfire smoke have robbed us of the glorious summer days we've loved about living here and cost us money. Even worse, portions of the summer are now threats for some people to simply breathe. Such was not the case fifty years ago, when our family visited every summer. Asthma rates nationwide have more than doubled since 1980 (Redd, 2002).

Psychological Impacts

Although much less commonly known, climate change impacts mental health as well. Given the mind-body connection this shouldn't be a big surprise. What affects the physical body affects the emotional body.

The American Psychological Association (APA) states that "PTSD, depression, general anxiety and suicide all tend to increase after a disaster." Also, alcohol and drug abuse, domestic violence and child abuse escalate (Clayton, Manning, Krygsman, et al., 2017). The effects of a destabilized climate compound the problem.

Drought, extreme heat, poor air quality, and other exposures are linked to suicide, particularly among people whose way of life is tied to natural resources (National Center for Health Statistics,

Centers for Disease Control & Prevention). Suicide is especially dangerous in Idaho. In 2020 the Gem State had the fifth highest rate of suicide in the nation, one and a half times the national rate (Garnett & Curtin, 2023).

The least known consequences of climate change, however, may be those foisted upon our children.

> "The climate crisis is the single biggest driver of health for every child born today."
> Lisa Patel, MD, MESc,
> Board member, Our Children's Trust,
> Executive Director, Medical Society
> Consortium on Climate and Health,
> Clinical Associate Professor,
> Stanford Children's Health

Dr. Lise Van Susteren is a distinguished psychiatrist with a special interest in the psychological impacts of climate change. An expert witness in both the *Juliana v. U.S.* and *Held v. State of Montana* climate change cases, Dr. Van Susteren states:

> Medical evidence shows children are uniquely vulnerable to psychological harms from climate change' (Declaration of Lise Van Susteren, paras 13, 28). 'Climate change is causing devastating physical impacts—injuries, illnesses, and deaths. But for the magnitude of its impacts, the potential insinuation into every aspect of our lives, the relentlessness of its nature and debilitating effects, it is the emotional toll of climate change that is even more catastrophic, especially for our children. ***It has the capacity to destroy children psychologically***" (Declaration of Lise Van Susteren, para. 12).

LOCAL CLIMATE CHANGE IMPACTS

To appreciate Dr. Susteren's concerns, it's important to understand that children are not small adults; they are distinctly different. Children's brains and lungs don't develop until their twenties. They have faster respiration and metabolic rates than adults, sweat less and spend more time outside (Expert witness testimony of Dr. Lori Bryant, June 2023).

In addition, children are particularly susceptible to extreme heat (Philipsborn & Chang, 2018). Infant mortality increases twenty-five percent on extremely hot days, with the first seven days of life representing a period of critical vulnerability (Basagana, et al., 2011). Heat is a leading cause of death and illness in high school athletes as well; nearly 10,000 episodes occur annually (Gilchrist, et al., 2010).

Contending with the physical impacts of climate change (e.g., floods, wildfires, droughts, and extreme heat) is an ongoing psychological challenge, especially for children, Dr. Van Susteren noted. The physical devastation of climate change can engender the feeling of "loss and disconnection" from "place and identity"—a condition previously described as solastalgia and as troubling as chronic stress (Dodgen, et al., 2016).

Dr. Van Susteren's continuing comment is particularly relevant to this book's theme:

> When places we have come to know well are irreversibly damaged, we lose the comforting sense of the familiar, the anchoring sense of belonging. Our inner psychic world—a key component of our sense of identity—mirrors the alien state of the damaged physical world—when it is lost, we lose a part of ourselves" (Clayton, Manning, Krysgman, et al., 2017).

Dr. Van Susteren next describes a few additional psychological impacts of climate change, primarily on young people.

Heat Waves and Violence

As temperatures rise, so does aggression (Raj & McDougal, 2014). The increased acts of aggression include assaults, murders and suicides, especially violent suicide. One researcher predicts that from 2010 to 2099 warming will cause an additional 30,000 murders, 200,000 rapes, and 3.2 million burglaries worldwide (Ranson, 2014). People with pre-existing mental disorders are especially vulnerable to the impacts of heat waves.

Heat waves also increase mortality rates "a couple of percentage points." Hospital admissions also rise, as do the number of suicides and the possibility of relapsing into mania in bipolar disorder and depressive episodes in depression. Finally, there is a "significant correlation between high temperatures and severity of symptoms in schizophrenia" (Lambert, 2022).

Indirect and Vicarious Climate Change Impacts

Worrying about the unprecedented risk posed by climate change, what lies ahead for the individual, their children and future generations takes a heavy toll on an individual's well-being. It wears people down, sending some to the breaking point, Dr. Van Susteren states. Children are especially vulnerable.

In a 2007 survey of Australian children, "a quarter of children are so troubled by the state of the world that they honestly believe it will come to an end before they get older." Reports described children crying, worrying about what is happening to animals, having problems sleeping, and wondering why their parents cannot do more (Tucci, Mitchell & Goddard, 2007).

In the first known case of what's been called "climate change delusion," a depressed seventeen-year-old Australian boy was hospitalized for refusing to drink water for fear it would cause the death of millions of people caught in the drought-ridden country (Wolf & Salo, 2018).

Institutional Betrayal

When disasters are no longer experienced as natural, as "acts of god or nature," but as having arisen or been made worse because of human behavior, it's much tougher for people to recover, Dr. Van Susteren notes. Injuries that occur as the result of an intentional act, or acts that could have been avoided, are much harder to put behind us. These are more psychologically damaging than injuries that occur accidentally (Ames & Fiske, 2013).

When trusted and powerful institutions that people depend on (e.g., schools, church, or government) are implicated in causing harm, the trauma is exacerbated. This "institutional betrayal" occurs "when an institution causes harm to an individual who trusts or depends on that institution" (Smith and Freyd, 2014, p. 575).

The federal government is the powerful institution, for example, in the Juliana v. U.S. climate lawsuit. In Montana the state government became the betrayer, in the *Held v. State of Montana* youth constitutional climate trial.

Looking down the road, Van Susteren states that:

> … our progeny will know government officials knew for decades the harm was coming to them. Knowing that we did not value them enough to bother protecting them from harm, which is how they will interpret actions today, will foment not only anguish but a dangerous feeling of cynicism

and distrust, breeding deep and enduring hostility towards democratic institutions, and toward each other as survival becomes an issue. It may not be realistic to imagine a well-functioning civil society under these circumstances (Smith and Freyd, p. 575). Many children ... are deeply and personally traumatized seeing beaches they play on disappear, having places they hike, camp, and recreate destroyed by wildfires, witnessing verdant areas around their homes dry up or be inundated with flood waters...or being unable to swim in water because it is contaminated with heat-induced toxic algae blooms (Smith and Freyd, p. 575).

No one can feel at home under these conditions. For many of us, children epitomize home, and hope for the future. Threatened children mean threatened homes and a threatened future.

Will tourists stop visiting? Climate change and our economy

How much climate change impacts the economies of Coeur d'Alene and Kootenai County, as well as businesses located here, is difficult to quantify. Companies are unwilling to disclose how much their businesses have been impacted by wildfires, wildfire smoke, less snow and rising heat. Thankfully, Row Adventures, headquartered in the Lake City, is a welcome exception.

When I was unable to connect in person with Vice-President and General Manager, Jonah Grubb, he was willing to answer questions by email, after a full day of meetings away from town. Operating throughout the Northwest, with a short season, Row Adventures is particularly vulnerable to this region's changing climate. Grubb was willing to provide some examples.

"Milder" winters that mean less water in the rivers means "shorter seasons on some of our rivers like the Moyie and St. Joe and the whitewater section of the Spokane River," Grubb related. If his company "can't book a late August trip on the Middle Fork [of the Salmon River] because guests are worried about lower water and smoke, that's a loss of about $75,000 in potential revenue." That would be about ten percent of the guests they regularly take on that river, Grubb estimated, acknowledging "the potential for significant loss."

Two trips on the Rogue River in Oregon had to be cancelled two years ago when the river was closed due to fire. The cost: $70,000 in revenue.

Wildfire smoke, though, presents "the most obvious and drastic changes in our operations. Views are diminished for guests, breathing is less pleasant, few stars to view while camping, etc." Some have respiratory concerns as well. As a result, Row Adventures has seen "diminished bookings in August due to guests being aware of wildfire smoke." More cancellations as well, some last minute, due to the smokey conditions (Grubb email, April 4, 2024).

Although the impacts on tourism and recreation may be less quantifiable with other local businesses, one thing is certain — the outdoors is key to our area's appeal and economic well-being. More than 10,000 county residents make their living in tourism's associated services, making it the county's single biggest employer. Kootenai County's employment in the tourism sector is forty percent higher than the national average (Legacy Industry Snapshot 10/21—Tourism Industry).

Similar to other states, the number one reason visitors come to Idaho is to visit friends and relatives. Being outdoors is number two in North Idaho. Four of the top six visitor activities involve being outside: hiking/backpacking, visiting landmark and historical sites, swimming and visiting national and state parks. Eight of the top ten visits center around water use. Going to Coeur d'Alene Lake and Lake Pend Oreille are the top two destinations (Borud, 2022).

Kootenai County is the number one boating destination in Idaho. County boat owners spend more money related to power boats (almost $90 million a year) than any other Idaho county (Borud, 2022).

While the impacts of climate change on local tourism and outdoor recreation are difficult to calculate, it's obvious the number of tourists hiking Tubbs Hill is less when it's shrouded in eye-burning smoke. The same for tourists wanting to fish or swim when the water is too low, too warm or where there are toxic algae blooms. Winter sports suffer when the snow is minimal or of poor quality, or when the ice is too thin to snowmobile on or ice fish in.

To date, though, neither wildfire smoke or rising heat have dampened reservations at North Idaho State Parks, particularly at water-adjacent campgrounds. Priest Lake State Park, for example, continues to be fully booked throughout the summer reservation season. Rising summer temperatures, therefore, will only exacerbate the demand for water-adjacent campgrounds, already in low supply (Idaho State Parks & Recreation study, 2017). As a result, overcrowding is a bigger concern to North Idaho state parks than climate change.

Will tourists go elsewhere if wildfire smoke worsens or heat waves become more frequent and last longer? Most likely, or stay home. No one knows for sure. What we do know, though, is that climate change makes our area's weather-influenced economy more vulnerable.

A final economic consequence of climate change: the expense of putting out wildfires. Prior to 2010, fire suppression costs for the Idaho

Dept. of Lands rarely exceeded twenty million dollars. Since 2012, costs have exceeded twenty million dollars and were over seventy-five million in 2015 and 2021 ("Overview of Fires in Idaho, 2015).

Climate-Affected Community Events

Back in 1970, the annual high temperature in the Lake City was just over 56°F. In 2017 it was just over 58°F, a two-degree increase. In 2015, though, it was a whopping 62°F. That year, the annual Art on the Green art festival was a scorcher. Was attendance affected by the brutal heat? I asked its organizers.

Art on the Green has been held from July 31st to August 2nd, on the campus of North Idaho College, since 1968. Fortunately, much of the festival is held under the shade of trees. This has likely kept attendance from dropping under higher temperatures, according to Carol Stacey, a longtime Board member of the Citizens' Council for the Arts, organizer of the festival.

However, there are no official attendance figures, says Anne Solomon, secretary of the arts group. Misters have been on the grounds for some time, she said, and she's noticed more and more people congregate under the forty-foot x forty-foot beer and wine tent, to enjoy the shade. A second tent was recently added and filtered water is available at water stations. In 2019, a benefactor paid to have a second water station added. All these improvements are related to rising temperatures.

Other annual events at risk from higher temperatures include the Ironman Triathlon in late June, the Coeur d'Alene Marathon in August and the North Idaho Fair, late August to September. All have fewer, or no shade trees, although the events have water, aid stations, and medical personnel on hand. Also, people can go indoors for certain Fair exhibits.

The late June-early July 2021 heat wave presented a major challenge to the 2,100 participants in the Coeur d'Alene Ironman Triathlon. Held on June 27th, under 100°F + high temperatures, approximately 600 competitors didn't finish. Twenty-eight percent of the participants did not finish, making it the second highest in the history of the race (Tiernan, 2021).

Finally, climate change involves a rarely acknowledged threat to public safety.

THE HIDDEN COST OF CLIMATE CHANGE

Heat and Violent Crime

As noted earlier, when temperatures soar, we become irritable, impatient and short-tempered. Some of us become aggressive as well, which can lead to criminal behavior.

A 2019 study in Los Angeles, for example, noted that "On average, overall crime increased by 2.2 percent and violent crime by 5.7 percent on days with maximum daily temperatures above 85°F, compared to days below that threshold" (Heilman, Kahn & Teng, 1). Violent crimes include murder and nonnegligent manslaughter, rape, aggravated assault and robbery. Aggravated assaults are attacks meant to inflict severe bodily injury.

In Idaho, seventy-five percent of violent crimes are aggravated assault; rape is about fifteen percent. Less than ten percent are robbery, murder and nonnegligent manslaughter, plus sexual assault with an object.

A second category, crimes against persons, involve force or intimidation, sometimes with a weapon. Simple assaults account for more than half of such cases statewide, aggravated assaults some twenty percent, intimidation and rape less than fifteen percent.

LOCAL CLIMATE CHANGE IMPACTS

What about locally, in Coeur d'Alene and Kootenai County? Do violent crimes and crimes against persons spike when temperatures jump? Idaho State Patrol records, from 2010 through 2023, for the Coeur d'Alene police department and Kootenai County Sheriff's Office were researched. Arrests made during the summer months, June through August, were compared to the rest of the year. If arrests during those months were in the top three for the year, that year was counted.

The results for the Lake City's police department: arrests for violent crimes, aggravated assaults, robbery and intimidation were made during ten of the fourteen summers. Disorderly conduct, which disturbs the peace, morals, or safety of the general public, appears directly related to elevated summer temperatures. Rape and simple assaults, on the other hand, are more year-round offenses. Too few murder and nonnegligent manslaughter arrests were made to research.

Kootenai County Sheriff's Office records show similar results. Arrests for violent crimes, aggravated assaults, disorderly conduct, robbery and rape were highest in the summer ten of the fourteen years.

Now for the bigger question: did these summer-focused arrests spike during times of high heat in the Lake City? Yes, during the following years:

- 2015 - Arrests for violent crimes and aggravated assaults were highest in June and August and for disorderly conduct in August.
- 2017 - June had the highest number of arrests for violent crimes and the second highest for aggravated assaults. August had the most disorderly conduct arrests.

- 2021 - August had the second highest number of violent crimes and aggravated assaults. Disorderly conduct arrests were highest in June and July.

The same pattern of spiking arrests during those hot summers existed in Kootenai County.

What does all of this tell us? Hotter temperatures from climate change threaten our public safety here at home, not just in big cities. That eroding sense of safety only adds to our felt loss of home.

Water & Climate Change

>"Water is the Life of All of Us."
>
>Felix Aripa,
>Coeur d'Alene Tribal Elder

Climate change has aptly been called water change. When temperatures rise, the quantities and forms of water change. Generally, snow turns to rain above thirty-two degrees F. and lake ice thaws. Persistent warmer temperatures dry the land to drought, wildfires worsen, and evaporation increases, putting more water in the atmosphere. This fuels larger storms, floods and simultaneously drops lake and river levels. Agriculture, recreation and power production suffer.

We've seen these change during the three-degree F average annual temperature increase in Coeur d'Alene since 1893 (Figure 3). Rising lake, stream and river temperatures have stressed or killed fish. Elevated air temperatures put more moisture into the air, intensifying rainfall. This increases sediment, nutrient loading (nitrogen and phosphorus from fertilizers), pesticides and herbicide runoff into rivers and streams. That pollution is carried downstream to lakes, leading to harmful algae blooms.

Here's another way to understand our three-degree temperature increase since 1893. Consider what happens to us when our core body temperature (e.g., 98.6°F) increases by three degrees. We become dehydrated and sick. What if our temperature didn't return to normal after taking antibiotics? What would happen if our temperature continued to climb? We would be in serious, life-threatening trouble. That's analogous to the situation the planet and all forms of life face from sustained overheating.

Now, what are the environmental impacts on local streams?

Studies show the frequency of large and deep pools of water in streams have decreased significantly in the Inland Northwest since the 1930s (McIntosh & Thurow, 2000). These pools provide rearing habitat for juvenile fish, resting habitat for adult fish as well as refuge from natural disturbances such as drought, fire and winter-icing.

Why are there fewer pools? Watersheds have been managed predominantly for extraction of resources, such as timber harvest, livestock grazing and mining. As such, ninety percent of the streams have roads along the banks or wetlands adjacent to the streams (McIntosh & Thurow, 2000). As a result, stream flows shrink or dry up, threatening fish populations and all aquatic life on the river banks. Tree cover diminishes as well, raising stream temperatures.

Climate change also threatens fishing, as noted previously, due to warming water, low water levels, flooding, wildfires and invasive species. In addition, declining snowpack, earlier runoff and extreme drought all reduce stream flows and lake levels, affecting fish of all types.

I called a forester in Sandpoint to find out more about our local habitat. William Love worked with the Idaho Dept. of Lands for thirty-three years, the U.S. Forest Service for five years and currently is a part-time consultant for Inland Forest Management. He's also the Panhandle Region representative for Trout Unlimited, a national non-profit dedicated to the conservation of freshwater streams, rivers and habitat for trout and salmon.

The Idaho Forest Group is headquartered in Coeur d'Alene and owns five lumber mills. In 2014, it purchased nearly 2,000 acres of Prichard Creek and nearby streams and forestland, east of Coeur d'Alene, Love said. The purpose was to make the area a conservation easement. This is a voluntary, legal agreement which permanently limits use of the land in order to protect its conservation value.

In that area, Prichard Creek flows from the Montana border through the historic mining town of Murray, into the North Fork of the Coeur d'Alene River. Gold was discovered in the North Fork near Murray in 1881, silver the following year.

Dredging thousands of tons of stream beds left large piles of rocks along the bordering banks. Those piles are being cleaned up, and fish habitat for West slope Cutthroat Trout increased, as part of the Idaho Forest Group restoration project.

A partnership between Idaho Dept. of Fish & Game and Avista Utilities is restoring the fish habitat for bull trout in Lake Pend Oreille as well. Logging dramatically reduced the "pretty robust" numbers of bull trout about 100 years ago, causing them to be listed as "threatened" on the Endangered Species Act. With their numbers recovering nicely, though, Fish & Game was petitioning to de-list the species in Lake Pend Oreille.

To protect cool water fish populations from warm water fish like small mouth bass (which decimate juvenile yolk sacs in a stream) and predator species like northern pike or walleye, Love described the need for what he called the 4 Cs in aquatic ecosystems: clean, connected, complex and cold.

Waterways need to be kept *clean* by keeping sediment and nutrient levels (nitrogen and phosphorous from fertilizers) down; *connected* by removing man-made barriers to fish migration such as dams and culverts; and *complex* by making sure there is woody debris and logs for the fish.

The first three Cs are directly influenced by compliance with the Idaho Forest Protection Act.

But cold is not, *because of the increasing warming influence of climate change, which is a "much larger hurdle,"* Love stressed (Dec. 2021 telephone conversation with Love).

What helps the rivers and streams stay cold? Reforestation—restoring mature native forest cover to shade rivers and streams, Love answered. In fact, he noted, restoring native trees along river banks would help to lower the water temperature a startling ten to twelve degrees F.

Forests & Climate Change

- Trees and their loss are critical to understanding climate change in the Inland Northwest and across the planet. In addition to taking carbon dioxide out of the air and putting oxygen back in, trees provide an incredible array of services. Consider how integral trees are to minimizing climate change:
- Trees draw large amounts of CO_2 from the atmosphere and put it down into the soil. From there it will leach into groundwater, which keeps it from quickly escaping into the atmosphere. Intact, old-growth forests, hundreds to thousands of years old, store the most carbon from the CO_2 molecules.

- A fully grown pine tree has a leaf area from a quarter-acre to over three acres, depending on the species. Rainforest trees have leaf surface areas as high as forty acres per tree. Trees literally breathe in the CO_2 through those leaf areas after we exhale it as biological waste, and then exhale oxygen as their own waste. Trees are our planet's lungs (Hartmann, 2004, p. 44).
- Rainforest trees will draw three million gallons of water up through their roots and release it as water vapor during its lifetime. Trees actually draw water *into* the soil. Without forest land pumping millions of tons of water into the atmosphere, there's little moisture released into the air to condense into clouds and then fall again as rain (Hartmann, pp. 44-45).

The roots of trees form new topsoil (over about 400 years, on average, to create a foot of topsoil capable of sustaining crops). Their deep roots break up lower levels of rocks, slowly bringing them to the surface, and shallow roots break up surface rock. Trees also draw minerals up into the tree itself to help make the plant matter (Hartmann, p. 49).

This is a brief description of the connection between trees and the need to preserve them to fight climate change. Important forest-related solutions, and a recent Biden administration action, will be described in the solutions section.

Climate Change's Impacts on Wildlife, Plants and Animals

In August 2022 the Idaho Department of Fish and Game published a draft *Idaho State Wildlife Action Plan*, to protect and conserve at-risk fish, wildlife and plants, and their habitats. For the first time, plants and a section on climate change were included in the plan.

LOCAL CLIMATE CHANGE IMPACTS

The plan remains a draft until officially approved by the U.S. Fish and Wildlife Service Director.

The report notes Idaho's annual mean temperature has increased 1.8°F since 1895. More heat waves, prolonged droughts, increasing spring and winter precipitation, but less consistent and earlier melting snow result. In addition, the report notes soil and fuel moisture are decreasing, causing increased wildfires.

Annual stream flow has decreased as well, with streams about 1.5°F warmer, with peak stream flows one to two weeks earlier. The plan predicts stream flows will continue to decrease and peak springtime stream flows could eventually be four to nine weeks earlier (Idaho Dept. of Fish & Game, 2022, 11).

The *Wildlife Action Plan* states that the Idaho Panhandle National Forest is in danger of being invaded by non-native weeds. Pollinators, whose numbers have dwindled for several years, are being threatened by climate change.

Relocating beavers to build dams in the wetlands would help retain water to improve conditions. This is an idea that may be considered (Idaho Dept. of Fish & Game, pp. 18, 225).

The following Idaho species in this area were listed as "potentially influenced by climate-related stressors": Mountain Goat, Moose, Fisher (weasel), Wolverine, Great Gray Owl, Western Bumble Bee, and White Bark Pine.

Finally, the *Wildlife Plan* indicated climate change is "potentially influencing" these local Idaho species: Chinook and Sockeye Salmon, Steelhead, Bull Trout, Yellowstone and Bonneville Cutthroat Trout, White Sturgeon, Moose, and Common Loon (Idaho Dept. of Fish & Game, pp. 18, 225).

Only voluntary recommendations are suggested to help conserve these species. More rigorous actions will likely be needed.

Losing Nature's Music

The Audubon Society's website describes how climate change impacts our local bird populations: "Rising temperatures and shifting weather patterns affect birds' ability to find food and reproduce...." (https://www.audubon.org/climate/survivalbydegrees/county?zipCode=83814).

Summarized below are the number of bird species expected to be vulnerable to climate change in Kootenai County over three warming scenarios — 1.5°C, 2.0°C and 3.0°C — during the summer.

> At 1.5°C warming — four species high vulnerability; fifty-nine moderate and fifty-seven low vulnerability.

The trumpeter swan and Franklin's Gull, for example, would be highly vulnerable by 2040.

> At 2.0°C warming — eleven species high vulnerability, seventy moderate and forty low vulnerability.

> At 3.0°C warming — forty-two species high vulnerability, fifty-two at moderate and thirty-six low vulnerability.

The Mountain Bluebird, the website noted, could lose half of its current breeding range and face drought and wildfire at 3.0°C warming.

As expected, the number of vulnerable bird species in Kootenai County rises with the temperature. For many of us the loss of their beautiful music and telltale of the seasons would signal a further erosion of home.

Now that we've discussed local impacts of planet-warming gases, there's another aspect of home to address.

Home as Social Connections

Home is a gift of a stable climate, but it also represents our social connections: our relationships with family, friends, neighbors, co-workers, and the greater community. We are formed through those relationships. Unfortunately, with climate change, the richness of those connections is being lost, as well as the stability of our home's climate.

It's no coincidence that the loss of both worsened with the rise of the automobile and its primacy.

> Once upon a time, in living memory of our grandparents, every city in America was a 15-minute city. Sometime after the war, we got the idea that cities were about highways and cars, and people had to "make way." But cities are for people, not cars. When you're getting your needs met walking, the air is healthier, people are healthier, people know each other ... people feel safer and they're more involved.
>
> Daniel Weir, Arlington County Planning Commission member (DeVoe, 2022)

A 15-minute city is about transportation, having what we need and want within a fifteen- to twenty-minute walk or bike ride. It has nothing to do with surveillance or control, as alleged by some.

A few years ago, while working at a counseling center in Hayden, I told the young receptionist that I was going to walk to the Super 1 grocery store to make a small purchase. The store was less than two blocks away. She asked: "What, aren't you going to drive there?" Surprised that anyone would need to take a car for that short errand,

I replied, a bit annoyed: "No, it's only across Highway 95, a couple of blocks away. I don't need to take a car for that!"

Having grown up biking or walking, it was unfathomable to me to have to drive two blocks. My brother and I walked the mile from our house in Ellensburg, Washington to the high school and back. We didn't get a car until we graduated in 1970. During family summer visits, we enjoyed walking, along with our sister, to see Grandpa at his downtown feed store a mile away.

Car Dependency: Its Toll

Today, however, the car is king. It seems like everyone drives everywhere, no matter the distance. This despite that more than half of U.S. trips are less than three miles; twenty-eight percent are less than one mile (FOTW #1230, 2022). All of our driving, however, has devastated the social connections at the heart of our towns and cities.

We're vaguely aware of some of the negative consequences of car dependency. The number of deaths on U.S. roads, for example — 42,795 in 2022, according to the National Highway Traffic Safety Administration (NHTSA, 2023). Then there's the financial expense of car ownership — $12,295 for the average U.S. household in 2022, according to the U.S. Department of Transportation. That's more than any other expense besides housing, with most of it going to cars (Wesseler, 2023).

Hidden from our view, however, is the social isolation car dependency imposes. We drive to work and to shop, isolated in our cars, often alone. In place of people we know, we encounter strangers at the bank, gas station, check-out counters, and more and more by mechanized transactions. We no longer walk to our local establishments, where we meet and greet one another, and merchants we know.

Without walking, we no longer know our neighbors. Sidewalks, too often vacant, are no longer enlivened by children playing. Riding bicycles through the neighborhood, as we did with friends growing up, is less and less frequent.

Also, some sixty percent of Americans report feeling lonely on a pretty regular basis, according to Laurie Santos, cognitive scientist and psychology professor at Yale University. Even college students, living in dorms with other students, report loneliness rates approaching sixty percent (PBS News Weekend, 2023). There's been a four-fold increase in the number of people who report having no close friends as well (Brooks, 2023).

In addition, loneliness results in serious medical problems. According to research cited by the Centers for Disease Control and Prevention, social isolation increases the risk of heart disease by twenty-nine percent and dementia by fifty percent. Isolation increases risk of premature death as much as smoking fifteen cigarettes a day (Wen, 2023).

Car-dependent neighborhoods often result in inactive lifestyles and obesity as well, which are responsible for up to ninety-five percent of diabetes cases in the U.S. (Wesseler, 2023). Both diabetes and asthma, other frequent consequences of low walkability neighborhoods, are extremely costly to treat, personally as well as for businesses and governments.

Another cost to our car dependency? Air pollution from tailpipe exhaust causes 200,000 annual "premature deaths" in the U.S. One M.I.T. study found that the leading cause of these deaths was vehicle emissions (Caiazzo, 2013). Unlike a generation ago, most air pollution now comes not from factories but from driving (Lutz, 2010).

Our car-centric culture contributes significantly to these physical, social and mental health costs. Through deliberate design, our cities prioritize moving cars, not people. Seen as the only way to get around, decisions made serve the auto, not what is best for us individually and socially. Common destinations have become too far to walk, forcing us to drive, isolating us from people we'd otherwise meet while strolling around the neighborhood. The experience of home as social connection in our car-centric culture is too seldom present.

Home is being lost.

PART III

RISING TO MEET THE CHALLENGE

"The effects we are seeing of climate change are the crisis of our generation."

DeAnne Criswell
FEMA Administrator

"To each generation comes its allotted task; and no generation is to be excused for failure to perform that task."

Theodore Roosevelt: Messages and Speeches

CHAPTER 9

Common Sense Climate Change

WEATHER IS THE CONDITION of the atmosphere at a particular time and place: e.g., rainy, hot, foggy, snowing, humid or dry, or clear. Weather has been likened to a person's skin temperature: it's volatile, can change drastically over a short period of time. Think how quickly our skin temperature changes in very cold or hot weather, yet with generally few ill effects (Henning, 2015, p. 29-30).

Climate, by contrast, is long-term. It's average weather for a particular area over a long period of time, usually measured over thirty years. It's akin to a person's core body temperature (i.e., 98.6° F), which cannot fluctuate much for us to remain healthy (Henning, 2015, p. 29-30).

The following analogy is helpful as well. Climate is what clothes you have in your dresser drawers; weather is what you wear today. Will Rogers summarized nicely, as only he could, the difference between climate and weather: "Climate is what you expect. Weather is what you get."

Finally, what influence does climate change have on El Ninos and La Ninas, those worldwide, naturally occurring and hugely influential weather patterns? El Ninos are hotter than average sea-surface temperatures across the tropical Pacific Ocean; La Ninas are cooler than average.

Recent research answers that question. Since about 1960, human-induced climate change has made both El Ninos and La Ninas stronger and more frequent. El Ninos result in hotter and drier summers and less snowy winters here; La Ninas the opposite (wetter and cooler). Here's the scoop on the latest research.

Over five years, a team of international scientists analyzed forty-three climate models, or computer simulations of Earth's climate system, establishing the stronger and more frequent action of both El Ninos and La Ninas since about 1960. They then compared climate simulations over hundreds of years before humans started ramping up greenhouse gas emissions after 1960. This analysis confirmed their calculations.

Finally, even if emissions are slashed to keep temperatures to 1.5°C (2.7°F), in line with the Paris Agreement goals, we can expect more frequent and intense El Nino events for another century. Why? Because the Pacific Ocean holds a lot of heat, which will take several decades to dissipate (The Conversation, 2023). We're in for a long ride of dangerous weather here: more heat waves, wildfires, smoke and droughts. Across the rest of the country, more hurricanes, tornadoes, and floods as well.

Now to Common Sense Climate Change, written by Craig Cooper, who has extensive knowledge of climate change. He notes that climate change often seems like an impossible science fiction tale. How could humans somehow mysteriously destroy the planet by doing millions of things each day, things we need to do to survive? How can carbon dioxide, CO_2, a naturally occurring, invisible gas that's a tiny part of our atmosphere be so dangerous? Well, by itself, it's not, he says.

But that's just the thing. CO_2 isn't alone. It's part of a larger system that depends on CO_2 staying within certain levels, to function reliably. Go outside of those boundaries, get too high, or change too fast, then the system gets out of whack. There are examples of similar things in our own bodies that need to be in balance for us to remain healthy and active.

For example, blood iron levels, thyroid activity, or even simply how much water we drink in a day. Keep these factors within certain boundaries and you can stay healthy. Stray outside those boundaries and you get sick and can die.

Like a human body, the Earth's climate system is relatively straightforward to understand in concept, and very complex when examined in fine detail. CO_2 is one of the main atmospheric gases that keep our planet at a livable temperature. Other gases include water vapor, methane, nitrous oxide, and a number of artificial gases only made by humans. Without any CO_2, the Earth's surface would be freezing, unable to sustain life. Most of the sun's heat that reaches the Earth would just reflect off the planet and escape back into space.

With CO_2, the Earth has a chemical blanket that captures a large amount of this heat and reflects it back towards Earth. This re-directed heat then moderates how much water vapor is in the atmosphere, providing another layer of insulation. Add more CO_2 and you add more heat and warm the planet. Take away CO_2 and you cool the planet. Just like adding and removing layers of clothes to maintain your own body at a healthy temperature.

This global process of temperature moderation by greenhouse gases has been known since the 1860s, when the world still ran on coal, and oil was first being discovered in America. This is not new knowledge. This discovery predates cars, airplanes, telephones, plastic, antibiotics, and pretty much everything that's "normal" to us today.

It's also true that the Earth has gone through natural periods of global warming and cooling throughout its history. These shifts had large impacts on our planet's ability to sustain life. The difference this time is that humans are farming CO_2. We are digging up carbon from deep in the ground (in the form of coal, oil and natural gas), burning it, and then chucking billions of tons of it into the atmosphere each year. About thirty-six billion tons per year. That's billion, with a B! This extra CO_2 isn't part of the planet's natural processes.

You can't dump thirty-six billion extra tons of a heat-absorbing gas into the atmosphere each year and expect it to not do anything. The Earth is a big place, but billions of extra tons each year adds up. Our atmosphere is not an indefinitely large trash dump. It can get over full. And when it does, bad things happen.

When our atmosphere gets over full with greenhouse gases, the planet warms up. Weather patterns change, ice melts, seas rise, snows shift to rain, droughts and floods become more common, crops fail more often and fires burn more severely. Get too hot, and large swaths of the planet can become too hot to sustain life. Coastal oceans become too hot and acidic to sustain their traditional fisheries. Disease spreads more easily.

People's lands become less productive, there isn't enough food, and millions of people are forced to migrate elsewhere to survive. These types of changes destabilize society and make us all poorer and less secure.

This is scary stuff. So scary that it seems almost impossible. But it's not. Work through the basic science, apply some common sense, and you can't help but conclude that too much CO_2 added to the atmosphere too quickly is a dangerous thing for us all.

We can argue about the details of how to predict the pace of change and its impacts. But we can't reject that it's happening, and that our explosive consumption of fossil fuels is driving us towards a deadly cliff. If you're driving a car towards a cliff, you should slow down and turn around. You don't accelerate and just hope for some miracle to save you.

Unfortunately, that's what our global society is doing. We're burning more fossil fuels each year and mostly hoping for a miracle. Well, that miracle hasn't yet occurred, and we're beginning to see the effects of the change that's already occurred. And the pace of change is quickening.

Droughts and forest fires are becoming more common. Mass migration of people from countries with more marginal land is happening, causing border crises around the world. The amount of ice-free sea in the Arctic is growing, glaciers are shrinking, and our winter snow is melting earlier. It's happening. It's time to put on the brakes.

Realities of Global Warming[2]

> ...imagine 400,000 Olympic-sized pools [nearly twice the length of Coeur d'Alene's Kroc Center lap pool—164' long, 82' wide, 6' deep]. That's about a billion metric tons of water, or 1 gigaton. Now multiply that by 36, yielding 14,400,000 pools. Thirty-six billion tons is the amount of [heat- trapping] carbon dioxide emitted in 2016 [worldwide] (Hawken, *Drawdown*, p. xiv).

---

[2] For simplicity, I'll use climate change and global warming interchangeably. Technically, they're different. Global warming is the rise in global temperature due mainly to increasing concentrations of greenhouse gases in the atmosphere. Climate change is the increasing changes in the climate over longer periods, including in precipitation, temperature and wind patterns.

COMMON SENSE CLIMATE CHANGE

In 2022, over 36.8 gigatons of carbon dioxide was emitted (International Energy Agency, 2023). In addition to that huge volume of carbon dioxide emitted every year, it takes a very long time to be absorbed back into the atmosphere—300 to 1,000 years (Buis, 2019). Approximately one-fifth of the carbon dioxide will still be around in 10,000 years (Gates, 2021, p. 18).

One final way to appreciate the threat of global climate pollution: "...we're trapping, each day, the heat equivalent of 500,000 Hiroshima-sized bombs" (McKibben, 2022).

As noted earlier in this chapter, balance is important in both human and ecological systems. Balance implies stability in a healthy system. Aware of it or not, we all depend on that stability. Unexpected weather, especially if it's prolonged, rattles us to our emotional core.

When wildfire smoke returned here in 2017, and years since, a scary change was signaled. Heat waves and other recent spikes in temperature did as well. Combined, they threaten not only what we love about our home but our lives and emotional stability.

Our entire civilization, in fact, depends on a stable climate. Science writer Fred Pearce noted this in a 2019 article:

> For some ten thousand years, climate on Earth has been stable. Remarkably stable. Since the end of the last ice age, we've spent four hundred generations taking advantage of this stability to build our civilization. Our massively complex society relies on the ability to plant crops knowing they'll grow, and build cities and infrastructure in places that won't be flooded by incoming tides or washed away by torrential rains (Pearce, 2019).

The climate system doesn't always change gradually either, Pearce added: "It takes big jumps based on tipping points." He cites the final few centuries of the last ice age, as ice sheets collapsed, sea levels rose twenty meters (sixty-five feet)—enough to drown much of Eastern England in less than 400 years.

Some 12,800 years ago, the world plunged into a 1,000-year deep freeze in a single year, with average temperatures crashing by sixteen degrees C (twenty-nine degrees F). Those were violent times, Pearce writes, and warns: "They could happen again" (Pearce, 2019).

One of the most threatening tipping points is the slowing of the complex network of ocean currents called the Atlantic meridional overturning circulation (AMOC). The AMOC, which includes the Gulf Stream, transports heat and salt through the global ocean, moderating the climate of Britain and northern Europe.

Writing in the journal *Science Advances*, scientists say the current has already declined by fifteen percent since 1950 and is on track toward an abrupt shift. That hasn't happened for more than 10,000 years. The current is being eroded by a faster-than-expected melt of Greenland's glaciers and the Arctic ice sheets in recent years. If the system collapses it would result in serious climate consequences from Europe to South America (Van Westen, et al., 2024).

How Do We Know We're Causing Climate Change?

For one thing, an overwhelming number of experts say so. By 2020, 198 scientific organizations worldwide had formally stated that climate change has been caused by humans (Governor's Office of Planning & Research, 2023). Eight additional points of evidence follow:

1. According to an Environmental Defense Fund article, **simple chemistry** reiterates human-caused climate change — when carbon-based materials are burned, carbon dioxide is emitted (based on research beginning in the 1900s).
2. **Basic accounting** of what's been burned, and therefore how much CO_2 we emit (data collection began in the 1970s).
3. **Measuring CO_2** and other greenhouse gases in the atmosphere, trapped in ice, show their levels increasing, higher than any seen in nearly a million years (measurements began in the 1950s).
4. **Chemical analysis** of the atmospheric CO_2 reveals the increase is coming from burning fossil fuels (research began in the 1950s).
5. **Basic physics** shows that CO_2 absorbs heat (research began in the 1820s).
6. **Monitoring climate** conditions find that the air, sea and land are warming, as we'd expect with rising greenhouse gas emissions, and ice is melting and sea levels rising (research began in the 1930s).
7. **Ruling out natural factors** that can influence climate like the sun and ocean cycles (research began in the 1830s).
8. **Employing computer models** to run experiments of natural vs. human-influenced simulations of Earth (research began in the 1960s).

What Are We Up Against with Climate Change?

Communicating the challenge that our community, the state, our country and indeed the world faces from climate change, without scaring people, is challenging. A logical place to start is with transportation, the biggest driver of greenhouse gas emissions in both Idaho and the country. Fifty-eight percent of Idaho's carbon dioxide emissions come from burning petroleum products in order to move people (Hall, 2019).

To reduce our country's climate warming pollution, the goal is to have two-thirds of new cars sold in this country be electric by 2032. How are we doing to meet a comparable goal here in Idaho? Of the more than 600,000 vehicles registered here, only 5,940 were electric, as of 2022 (Nasdaq, 2024). That's one percent.

However, we need it to be sixty-six percent within only eight years. To call that a challenge is a huge understatement. Even California, with the highest adoption rate of electric vehicles in the country, at forty-four percent, is behind the eight ball.

With transport, at least the technology and products are available. Not so with growing food and heavy industry (including manufacturing), which account for one-third of the country's carbon footprint. Progress towards decarbonization has been much slower in those technologies than in electric vehicles and wind and solar energy (McDonnell, *Semafor*, 2024).

In March 2024, the Dept. of Energy awarded grants totaling six billion dollars to kickstart dozens of heavy industry projects into commercial production. One project would use hydrogen, not coal, to make steel; another would make limestone cement using electricity. A third would capture carbon dioxide and store it in geologic formations (*PBS Newshour,* March 29, 2024).

While this funding represents an important first step, virtually every activity in our modern lives involves emitting heat-trapping gases. Meeting our basic needs without emitting planet-destroying greenhouses gases will be the biggest challenge humanity has ever faced.

CHAPTER 10

United We Stand

"A house divided against itself cannot stand."
Abraham Lincoln
1858

MUCH INFORMATION HAS BEEN presented so far, some of it uncomfortable, even scary. However, we can take comfort knowing that a number of us have faced and defeated even more life-threatening odds by coming together. My own family, for example, did years ago.

While working in Spokane during the summer of 2001, I got a mid-week phone call from my mother. My identical twin brother, Dick, was in a Boise hospital in some pain. She and Dad were headed down to see him the next day. Would I drive down with sister Janet to see him the coming weekend? Sure, I said. Although concerned, Mom didn't sound alarmed.

By the time Janet and I got there, though, the news was grim. Dick had been diagnosed with stage 4 non-Hodgkin's Lymphoma (NHL). X-rays showed cancer had spread throughout his entire lymph system. He looked like a scarecrow when we visited him in his hospital room. Emaciated and in a great deal of pain, a five-foot walk across the room was excruciating to watch.

Privately, the doctor told us that if he didn't begin chemotherapy immediately, he'd likely be dead within two weeks. I gulped. *That was scary news. But it needed to be said, and we needed to hear it, to develop a course of action.* It was a crisis, unfortunately, faced by many families. *That same scary truth needs to be said about the climate crisis, so we can all act accordingly.*

Chemo for Dick was started that night. Sitting soberly in the hospital dining room that evening, Mom, Dad, Janet and I developed a plan. We'd drive Dick to Coeur d'Alene, where he would continue his cancer treatments while living with our parents. The next day, Mom and I left in separate cars to get their home ready. After collecting Dick's things from his apartment, Janet, Dick and Dad would follow the next day in Dick's car.

Dick's prognosis was frightening: few survived NHL at the time. I spent much of the drive home in tears. The strong possibility of losing my identical twin brother was overwhelming. Born three minutes apart, he was Pete to my Repeat; pitcher to my catcher. We were doubles partners in high school and college tennis. We were (and remain) very close.

Between bouts of grieving, I wondered what I could do for him in his final days. A devoted Green Bay Packers fan like myself, I decided that I'd call the Packers organization to see what they could send him.

My brother's dire circumstances, however, brought an unforeseen development: my parents came together like they hadn't in years. In their mid-seventies, married for over fifty years, they'd seemed to lead such separate lives whenever I visited. They ate meals together in the kitchen, yet engaged in little but surface pleasantries. Sadly, there seemed to be little love shared between them.

But they came together to save my brother's life. A retired nurse, Mom tended to his port line and fixed his meals. She submitted paperwork for his disability claim. Dad joked and engaged in banter about the Packers with him, and helped to transport him to monthly cancer treatments. They encouraged and loved him. Along with advances in cancer treatment, thankfully, my brother recovered. He's been cancer free for over twenty years now and happily married for ten.

Clearly, not all family's cancer stories have such fortunate endings. Some lose children much younger than my brother. However, what's key in many cases is how families rally in the face of an immediate threat. The longest, scariest of odds, can be beaten when a family's love and the best of medicine combine.

There are some similarities here with the climate crisis—for example, the immediacy and personal, local impacts. Clearly though, the climate crisis is of a different order threat, one that threatens civilization itself. Some of the differences between cancer and climate change bear noting.

First, climate change has been obfuscated and lied about by the fossil fuel industry for decades, for their short-term economic gain. Gutless, and/or corrupt politicians have been bought off, and the public misled. Therefore, few of us know that burning fossil fuels causes nearly nine million premature deaths worldwide *every year*, more than have died from the pandemic in three years (Vohra, et al., 2021). The threat of cancer, on the other hand, is clear to us all.

Second, the climate crisis involves a loss of trust in institutions that have failed to protect us from the threat. Too many elected officials have valued money and political power over human life and long-term ecological health. It's that mistrust of government which causes opposition to government efforts to address the climate. Our own personal physicians, however, are still generally trusted.

Finally, global climate change is exacting an accelerating toll on more and more people's lives around the world. As noted earlier, the disrupted climate is displacing people's past, present and future lives, their stability, security and safety (Askland, et al., 2022). Cancer, fortunately, can increasingly be cured, or successfully treated. However scary, if it doesn't terminate a person's life, it doesn't eradicate one's being like climate change is increasingly doing.

We must come together to save ourselves, our loved ones and the homes in which we live, as my parents did, and so many others have done, before and since.

Public Perceptions of the Problem

Unfortunately, too many of us still can't agree that global warming is a problem. There's been some progress since I began writing this book in 2018. At that time little was said about global warming by the Idaho governor or local officials. The issue, even its existence, was ignored or minimized. I had little reason to think anything would change. As it turned out, though, I was soon mistaken.

In January 2019, newly sworn-in Republican governor Brad Little publicly stated "Climate change is real" (Russell, 2019). Finally, an acknowledgement of the truth. Then in May, Idaho Republican Rep. Mike Simpson acknowledged climate change and proposed to take action to fight it (Johnson, 2019).

Since 2018, the values of the American people toward global warming have been studied by researchers from Yale and George Mason University. They developed what's called "Global Warming's Six Americas." They describe the following six distinct populations of U.S. citizens, based on their views on global warming: the Alarmed, Concerned, Cautious, Disengaged, Doubtful and Dismissive.

The *Alarmed* say climate change is happening, it's human-caused, an urgent threat and they strongly support climate change policies. The *Concerned* agree climate change is a serious threat, support climate change policies, but believe climate change impacts are distant in time and space and therefore of a lower priority than the *Alarmed*. The *Cautious* have not made up their minds whether climate change is happening while the *Disengaged* know little about global warming and rarely or never hear about it.

The *Doubtful* don't think global warming is happening, or it's just a natural cycle and not a serious risk. The *Dismissive* say global warming isn't happening and they endorse conspiracy theories ("global warming is a hoax") (Leiserowitz, et al., 2023).

The latest Global Warming's Six Americas report, in December 2023, found that fifty-seven percent of Americans say they are at least "somewhat worried" about global warming. This includes twenty-eight percent of Americans who say they are "alarmed." In the past ten years, the Alarmed segment has nearly doubled in size (from fifteen percent to twenty-eight percent of the adult U.S. population). The Dismissive segment, on the other hand, has decreased from twenty-six percent in 2013 to fifteen percent in 2023 (Leiserowitz, et al., 2023).

A majority of Americans understand global warming is mostly human-caused. Only twenty-eight percent think it's caused mostly by natural changes in the environment. More than half of Americans think global warming will harm people in the U.S., people in their community, or their family. Seventy-one percent think it will harm future generations of people. Forty-seven percent of Americans also think they themselves will be harmed (Leiserowitz, et al., 2023).

There's also a specific survey of Americans' beliefs, risk perceptions and policy support regarding global warming. The Yale

Climate Opinion Maps show results for every county in the country. Kootenai County voters score below the national average in their views on global warming. The 2023 result showed sixty-four percent of the county residents say global warming is happening, vs. the national average of seventy-two percent.

Less than half say it's caused mostly by human activities (fifty-eight percent is the national average). Just over half are worried about global warming; sixty-four percent is the national average. Three-fourths of Kootenai County residents support funding research into renewable energy sources, compared to seventy-nine percent nationally (Marlon, et al., 2023).

CHAPTER 11

Whom To Trust (And not)

T HE SUBJECT OF CLIMATE change, most of us know, is incredibly divisive and fraught with highly emotional rhetoric and misinformation. Given this situation, whom can the reader trust with accurate information? One of those sources, I argue, is the military.

The Military

First of all, the military doesn't have a financial or political stake in climate change. Their mission is to defend the country, our Constitution and our national interests, both at home and abroad. What military leaders say publicly on this issue ought to be above reproach. In addition, many live and work in areas all over the world, so they have a breadth of experience few others can offer.

In April 2021 the U.S. Army Corps of Engineers reported on the Department of Defense's (DoD) installation exposure to climate change. In its Executive Summary, it noted:

> Climate change has been identified by the DoD as a critical national security threat and a threat multiplier.

> Improvements to master planning and to infrastructure planning and design are recognized as vital for reducing current and future vulnerability to climate hazards to installations, missions, and operations worldwide (Pinson, et al., 2021).

By mid-July 2021 the Department of Defense prepared a comprehensive "Climate Risk Analysis" — on the same timeline as a major review of the U.S. approach to China (Guy & Sikorsky, 2021). Such urgency indicates the level of risk climate change poses to the military.

Earlier, in January 2019, the DoD released a report on the effects of climate change on mission-critical bases around the country. It found "recurrent flooding, drought, and wildfires are the primary concerns at the 79 installations" (Dept. of Defense, 2019).

Former Navy secretary Ray Mabus put the military risks of climate change succinctly:

> Climate change is a national security risk. Stronger storms will lead to increasing damage to coastal military facilities, as when Hurricane Michael caused substantial damage to Tyndall Air Force Base in Florida [in Oct. 2018]. Stresses on resources and agricultural changes will increase the global flow of refugees and cause cross-border instability. That, in turn, will mean greater involvement of U.S. forces around the world (*Air Force Times*, 2018).

Of the Pentagon's facilities at risk from climate change, Naval Station Norfolk, the world's largest naval base and the headquarters of the U.S. Atlantic Fleet, has received the most news coverage.

WHOM TO TRUST (AND NOT)

Much of this critically important facility is already prone to flooding during high tides and storm surges. With climate change, the frequency and extent of the flooding will increase substantially.

By the end of the century, according to calculations by the Union of Concerned Scientists:

> "... 60 percent of the installation's land area will be exposed to tidal flooding on a regular basis, and 95 percent of it will experience ten feet or more of flooding during a Category 4 hurricane. Long before this, however, 'portions of NS Norfolk would be inundated with each high tide,' essentially rendering them useless" (Siegfried, et al., *Union of Concerned Scientists*, 2016).

Due to these worsening conditions, it's inevitable that this huge facility, along with several other major military facilities, including the Norfolk Naval Shipyard and Langley Air Force, in the Hampton Roads area of Virginia, will have to be moved. The cost will be massive and a threat to our country's national security when undertaken.

Climate change threatens our national security in other ways, according to the military. Rising seas, increasing temperatures, and increased drought will eventually displace the likely hundreds of millions of people who live in low-lying coastal areas or the burning deserts of the Middle East. For example, a severe drought in Syria from 2007-2010 forced farmers off their lands and helped trigger the Syrian conflict. Unstable nations across the Middle East are already facing water shortages that will only worsen as temperatures continue to rise.

The U.S. military doesn't say that climate change will cause wars. They call it a "threat multiplier." In places like Africa, where poverty is widespread, governance in numerous nations is poor, where much of the population relies on rain-fed subsistence agriculture, climate variations can be extreme. Adding the stress of a disrupted climate to that toxic brew has accelerated the existing threats and tipped fragile states toward war.

Conservative Values and Climate Change

Republicans in Congress have opposed climate change legislation for decades, one Republican says, because government regulations are contrary to their values. A former two-term U.S. Representative from South Carolina, Bob Inglis, is deeply conservative. However, in a 2010 primary to serve a seventh term, he committed the political heresy of stating climate change is real. He was beaten three to one in the primary.

After licking his wounds, he founded republicEn.org, a national group which touts free enterprise solutions to climate change. A project of the Center for Climate Change Communications at George Mason University, members of republicEn say they're part of the Eco Right, a balance to the Environmental Left.

Led by Inglis, republicEn endorses putting a price on carbon through carbon fee and dividend legislation (which will be discussed in chapter 13). Inglis is also a member of the advisory board of Citizens Climate Lobby, a group guiding that market-driven strategy through the U.S. Congress (described as well in chapter 13).

In a 2017 guest column in the *Greenville Journal* (S.C.) newspaper, Inglis, noted what republicEn.org believes in:

> ...a smaller government...the power of markets, accountable free enterprise, the power of price signals like the ones Milton Freeman taught, the liberty of enlightened

self-interest in the innovation that can come from a world of consumers seeking clean energy. We don't believe in the growth of government. At republicEn.org we want to help fellow free enterprisers step forward with solutions to climate change that bring more energy, more mobility, and more freedom. We're energy optimists and climate realists (Inglis, op-ed, 2017).

Faith and Climate Change

> The earth is the Lord's, and all its fullness,
> The world and those who dwell therein.
> For He has founded it upon the seas,
> And established it upon the waters.
>
> Psalm 24

Another trustworthy voice is the evangelical Christian and Nobel Prize-winning atmospheric scientist, speaker and author, Katharine Hayhoe. Hayhoe says there's a long history in evangelism of advocating for "creation care," the belief that God charged humanity with caring for the Earth.

Hayhoe has addressed hundreds of groups and thousands of individuals on climate change. In a September 2021 podcast on hurricanes and climate change, she indicated that "75% of Americans say climate change is real. But only 35% think it matters in their lives, a phenomenon called "psychological distance." This psychological defense allows people to think something is happening in a distant place, in some future time and to other people, not themselves.

Hayhoe stressed that the thirty-five percent figure is the number we need to increase, "to move the needle" on climate change (NOVA Now podcast, 2021). Moving that needle locally is a key motivator for writing this book, even before I'd heard of psychological distance.

153

In 2015 Pope Francis sent a papal letter to all bishops of the Roman Catholic Church. This climate encyclical was entitled Laudato Sí, or "Praise Be to You: On Care for Our Common Home." In the encyclical Pope Francis blames the climate crisis predominantly on human actions, and the inertia of government and business actions:

> To take up these responsibilities and the costs they entail politicians will inevitably clash with the mind-set of short-term gain and results which dominate present-day economics and politics.... But if they are courageous, they will attest to their God-given ability and leave behind a testimony of selfless responsibility. (Dias, 2018)

On March 9, 2023 Cardinal Michael Czerny, Pope Francis' primary voice on climate and a Gonzaga University graduate, spoke at an event in Spokane. It was hosted by Gonzaga's Center for Climate, Society and the Environment.

Long an advocate for human rights, Cardinal Czerny sees "caring for the climate means caring for our neighbors." Those neighbors include "Indigenous peoples and migrants whose homelands have deteriorated as a result of climate change."

"Nowadays, the problems facing the climate are common knowledge, so action is needed. The problem now isn't ignorance,' he said, 'the real problem instead is indifference and despair."

"In facing the climate crisis, 'Blaming God is not helpful,' he emphasized. 'Re-ask the question: Where am I and where are we in all this?' "The best antidote to despair and indifference is action, Czerny said, and dialogue is a vital part of that action. "'Dialogue and better politics are the only way out,' he said" (Hanlen, 2023).

Idaho Politicians Weigh In

As noted earlier, a few Idaho Republican politicians have acknowledged the existence of human-caused climate change and the need to do something about it. On January 16, 2019, the newly sworn-in Republican governor, Brad Little, "shocked some at the Idaho Environmental Forum by declaring that climate change is real and we will have to deal with it" (Russell, 2019).

Little stated, "Climate change is real. I'm old enough that I remember feeding cows all winter long in deep snow, and I go to the ranch now and say, 'You wimpy guys, back in the old days when I was a kid, we had winters.' And there are other things. The ecosystems are changing.

"'Little said even the microbes in the soil are changing, altering how they take in oxygen. So, when we do a range project, you've got to make that to where it's changing for a different climate type. The silviculturists will tell you, because of the changing climate, the number of trees, the mix of trees, it's all changing. You've gotta adapt to it. [Silviculturists grow and cultivate trees]

"Climate's changed, there is no question about it," Little said. "We've just gotta figure out how to cope with it and we gotta slow it down. Now, reversing it is going to be a big darn job, if 90 percent of the scientists are right.... The rate of change, we can make a change in. But it's a big deal" (Russell, 2019).

Another apparent political shift occurred later that year. In a May 2019 magazine article, Idaho Congressman, Republican Mike Simpson, acknowledged the existence of climate change and proposed taking action to fight it. "Climate change is a reality," he said at a conference in Boise. "It's not hard to figure out. Go look at your thermometer" (Johnson, N., 2019).

Capital Investments

What is the investment world doing to combat climate change? In the summer of 2018, Black Rock, the largest asset management company in the world, staged a stockholder rebellion among Exxon Mobil investors. It forced the company to report yearly to its stockholders the impact of global measures to limit global warming to two degrees C (3.6°F) over pre-industrial levels.

Black Rock investors wanted to know what would happen to the value of their stock if a large percentage of oil, coal and gas reserves are not allowed to be burned (called "stranded assets"), to meet worldwide climate change measures. Since then, nearly 150 other oil and gas companies have had to provide answers to those same questions, in yearly submissions to their stockholders.

May 26, 2021 saw shakeups in the boardrooms and practices of three multinational oil companies, all related to risks posed by climate change.

First, a Netherlands court required Royal Dutch Shell to reduce its climate emissions by forty-five percent by 2030 (compared to 2019 levels). This would align with what the Intergovernmental Panel on Climate Change said it would take to limit the planet's temperature increase to 1.5°C (2.7°F) by 2030. This was believed to be the first court case to rein in emissions, targeting a multinational company (*Christian Science Monitor*, 2021).

That same day, sixty-one percent of Chevron's shareholders voted in favor of a proposal from a Dutch environmental group, Follow This, to cut the company's carbon emissions. The group's founder, Mark van Baal, told *The Guardian* the shareholder coup represents a "paradigm shift" and "a victory in the fight against climate change.... Institutional investors understand that no investment is safe in a global economy wracked by devastating climate change," van Baal said (*The Hill*, 2021).

WHOM TO TRUST (AND NOT)

Later that same day, shareholders at Exxon Mobil forced the election of three climate activists onto its Board. The story sounded like David had felled Goliath. The shareholder revolt was led by a six-month-old hedge fund, Engine No. 1, which owned a meager 0.02% of Exxon Mobil, a $250 billion behemoth. Exxon's Achilles heel, though, was its loss of twenty-two billion dollars in 2020, and its returns consistently disappointing shareholders over the last ten years, according to Engine No. 1 (Subramonian, 2021).

The new hedge fund convinced other shareholders, such as the massive asset management firms BlackRock and Vanguard, which together own roughly fifteen percent of Exxon Mobil, to follow their lead to make important changes. The leader of the bold campaign for Engine No. 1, Charlie Penner, told Bloomberg: "If you can get Exxon to change, everyone else in the industry has to listen" (Subramonian, 2021).

That action paid off in an early win as well. Since placing the climate-focused independent directors on the Board in May 2021, Exxon Mobil scaled back its long-term production targets by twenty-five percent from forecasts before the pandemic started. That would be ExxonMobil's lowest level of oil output in two decades, through 2025.

The new directors are having the oil giant allocate capital and plan for growth by including the social cost of oil. "That is, how to price the future harm caused by release of one additional ton of CO_2, using present monetary value" (Carlson, 2021).

Three years later, however, that optimism has all but disappeared. In a February 2024 interview with *Fortune*, Exxon Mobil CEO Darren Woods said his company wasn't interested in building out wind and solar power because "we don't see the ability to generate above-average returns for our shareholders" (McKibben, 2024). In

157

other words, they'll stick with ruining the planet by burning their oil and gas reserves, rather than harnessing what the sun and wind deliver for free, to make their traditional returns.

What's even more galling, though, is that Woods claimed that the world "waited too long" to start developing renewables. This despite Exxon's scientists, knowing all about climate change forty years ago, forecasted the temperature in 2020 with remarkable accuracy. As a result, "they began to build their drilling rigs higher to compensate for the sea level rise they knew was coming, and plotting out which corner of the Arctic they would drill once it melted."

Instead of telling the truth it knew, however, it "erected a huge architecture of deceit and denial and misinformation" over the next three decades (McKibben, 2024). It's those lost decades, due primarily to Exxon Mobil's coverup, that has put the planet in today's climate crisis. Yet its CEO blames the world for "waiting too long" to develop renewables! What shameless deceit.

As if Exxon Mobil's arrogance wasn't enough, consider that companies such as Apple, Amazon and Microsoft are bankrolling new fossil fuel projects. How? According to a recent report, "the cash they kept in the bank ... got lent out to build more fossil fuel infrastructure." The report makes clear the consequences: "If the largest banks and asset managers in the U.S. were a country, they would be the third largest emitting country in the world, behind China and the U.S." (Carbon Bankroll 2.0, 2024).

In March 2024 the U.S. Securities and Exchange Commission, after nearly two years, announced that it would require corporations to inform investors of their greenhouse gas emissions. Good news, but the rules were much less rigorous than previously discussed. Gone was the need to disclose not just planet-warming emissions from their own operations but emissions all along a company's supply chain.

In addition, the biggest companies will only have to report the emissions they consider "material" to investors, leaving corporations like Exxon Mobil leeway. Thousands of smaller businesses are exempt as well, another big change from the original proposal.

Why these significant changes? Lobbying, primarily by the fossil fuel industry, argued that the regulations would be onerous and expensive. To survive legal challenges as well.

The bottom line about the corporate world and the climate threat? Be dubious about what they say or do. Money and political power in this country have an enormously corrupting influence.

Philanthropic Institutions

Recently, philanthropic institutions have announced diverting millions of dollars from fossil fuels, due to climate change concerns. In mid-October 2021, the Ford Foundation announced such a move. Both the MacArthur Foundation and Harvard University, in September 2021, announced their institutions would end investments in fossil fuel related companies ("Ford Foundation nixes fossil fuels," *The Spokesman-Review*, 2021).

The Insurance Industry

Why is the insurance industry considered trustworthy regarding global warming? Because it takes a longer-term, purely economic view of risk than other businesses and assesses risk in specific dollar amounts. They have to pay on claims they cover, so they have no financial incentive to inflate those climate-related costs.

How many "natural disasters" have occurred in the U.S. over the past ten years? According to the NOAA National Centers for Environmental Information (NCEI), 152 disasters caused at least one billion dollars of damage per occurrence. This puts the total cost of billion-dollar disasters to more than $1.1 trillion over the past ten years. *Climate change plays a major role in the frequency and intensity of severe weather* (Martin, 2023).

In 2023, there were $95 billion of insured losses in the U.S., primarily due to record thunderstorms (Munich RE, 2024).

That same year, five large U.S. property insurers, including Allstate, American Family, Nationwide, Erie Insurance Group and Berkshire Hathaway:

> told regulators that extreme weather patterns caused by climate change ... will [cause them to] cut out damage caused by hurricanes, wind and hail from policies along coastlines and in *wildfire country,* according to a survey conducted by the National Association of Insurance Commissioners (Gogage, *The Washington Post*, 2023).

Idaho is number five in states that have experienced the most loss from wildfire damage (Martin, 2023).

PART IV
SAVING HOME

"Being a parent means loving your children more than you've ever loved yourself."

CHAPTER 12

What You Can Do

"The reason we don't have slavery today, the reason women can vote, the reason the Civil Rights Act passed is because ordinary people decided the world had to change. We have to activate every single one of us."

Katharine Hayhoe, climate scientist at Texas Tech University, chief scientist for The Nature Conservancy, author of *Saving Us: A Climate Scientist's Case for Hope and Healing In a Divided World* (2021)

THERE ARE OPPORTUNITIES FOR each of us, in our everyday lives, to contribute to resolving global warming. They will all help to preserve our gifts of home. To do so, we need to keep worldwide warming no higher than 1.5°C (2.7°F) above pre-industrial times, as enshrined in the Paris agreement. In mid-2023, the world had already warmed 1.2°C (2.16°F). The news is not at all good if we don't.

A study undertaken by the re-insurance company Swiss Re, for example, estimated that just over double that warming — 2.6°C (4.68°F) warming by 2050 — would inflict three times more economic damage than the COVID-19 epidemic. "Unlike the COVID epidemic, however, the damage inflicted by warming will only grow worse every year" (Guo, et al., 2021).

WHAT YOU CAN DO

To keep temperatures below that 1.5°C threshold, the U.S. needs to halve its 2005 level of greenhouse gas emissions by 2030. Thankfully, in 2023, we're nearly halfway there, according to Columbia University's Dr. Melissa Lott, on the April 2023 NOVA program, "Chasing Carbon Zero" (Chasing Carbon Zero, 2023).

High income countries, like ours, have been reducing their CO_2 emissions over the past decade: the U.S. at a rate of 0.7% per year, the European Union 1.4% per year. Overall, high income countries have reached a comfortable standard of living (Friedlingstein, et al., 2022).

Middle and low-income countries are much different, though, with lower standards of living than in Europe and the U.S. To improve those living standards, "energy use is growing rapidly.... [Yet] Rapid growth in solar and wind power is not sufficient to capture the increased energy demand, and consequently fossil fuel use and CO_2 emissions continue to rise" (Friedlingstein, et al., 2022). Therefore, America's CO_2 emissions must drop by fifty percent by 2030, to meet the 1.5°C. target, Europe's by fifty-five percent.

What each of us does to manage what we consume (from the Latin *consumere*, to use up) plays a role in maintaining a livable planet. Consider: "More than 60 percent of greenhouse gas emissions and up to four-fifths of land, material and water use stem from household demand...." (Ivanova, et al., Summary, 2016).

Our demand for stuff, to entertain us, immediately fix what ails us, to have the latest and fastest, to have what someone else has, is peddled as needs by advertising. Therefore, we need to make a big dent in consumerism.

Consider the following examples of our rampant materialism: "The average American home has tripled in size in the past half century, though families have become smaller. A household in the U.S. contains, on average, 300,000 individual items—no wonder one in ten households rent a storage unit and one in four people with a garage say it is too full to house a car" (Springer, 2017).

In addition, we too often buy unconsciously. "According to CNBC, the average U.S. consumer spends $5,400 per year on impulse purchases" (Berners-Lee, 2022).

Cutting consumerism

> "I only feel angry when I see waste. When I see people throwing away things we could use."
> Mother Teresa
>
> "Waste not, want not."
> Benjamin Franklin

How can we cut consumerism? Develop a "1.5°C lifestyle," one commensurate with the year 2100's temperature boundary. What would that involve? "Phase out [of] private jets, mega-yachts, fossil-fueled cars, short flights and frequent flyer rewards. In their place, we'd need excellent rail networks and electric car-sharing schemes and dedicated bike and bus lanes" (Raworth, 2022). Plus, do we really need an expensive half ton pickup to drive to the grocery store, with all of its attendant environmental costs?

This brings us to a virtually sacrosanct notion in our culture: that our economy must grow (i.e., the Gross Domestic Product) if we're to survive. Some economists disagree.

An economic anthropologist from Spain, for example, says:

> High-income countries don't *need* more growth.... it is possible to meet human needs at a high standard with

WHAT YOU CAN DO

much less energy and resources than rich countries presently use. The key is to scale down less necessary forms of production and organize the economy around human well-being rather than capital accumulation. This is known as *degrowth*... (Millward-Hopkins, et al., 2020).

Recent news lends credence to the urgent need for degrowth. A team of Earth scientists reported that seven of the eight global boundaries of systems that provide a "safe operating space" for civilization have been dangerously surpassed by humans. These systems include the climate, biodiversity and fresh water (Rockstrom, et al., 2023).

Degrowth

The key to degrowth is to: "...decide which sectors of the economy we really *need* to improve (e.g., renewable energy, public transport, and health care), and which are clearly destructive and should be scaled down (SUVs, air travel, fast fashion, industrial beef, advertising, finance, the practice of planned obsolescence, the military-industrial complex, and so on)" (Newell, et al., 2021).

This raises the question, though: what about people's jobs? Since there would be a need for less labor, the economist notes, the work week could be shortened and work more evenly shared. In addition, public works programs could be developed to participate in most collective projects. For example, building renewable energy capacity, insulating homes, producing local food and regenerating ecosystems (Newell, et al., 2021).

Many who have had parents raised during the Great Depression, as I did, will have fewer struggles adopting a 1.5°C lifestyle than others. Dad barked at us if we left the light on in a room, showers were to be short and we didn't get dessert if we didn't clean our plates.

Today I continue that same frugality. I ride my bike or walk, rather than take the car, as often as possible. My significant other and I keep the heat down in the winter and we will sparingly use a recently purchased air conditioner. We only run the dishwasher, and wash clothes, when the loads are full. I haven't flown in years and when I do, I'll stay longer to experience each place and take fewer flights.

One final point—the "false dichotomy" between two extremes—individual actions and the systemic. Focusing on either extreme "overlooks the vast territory in between," one author states, called social influence. Climate action occurs through the many roles we all play in constant contact with one another—i.e., "… in the life of communities, families, friendship groups, organizations and work places" (Graziano & Gillingham, 2015).

For example, "On average, if two houses within a half-mile radius install a new [solar panel] system, the resulting peer influence prompts one additional household to do likewise (Seebauer, 2015). In a similar way, growth in the uptake of electric bikes, scooters and cars has been directly enabled by people discussing their use and encouraging others to try them" (Berton, 2019).

In Sweden, after the phenomenon of flygskam (flight shame, from Greta Thunberg's influence) took root, domestic air passenger numbers fell by nine percent between 2018 and 2019 (Wood, 2021). The power of social influence.

In addition, "Personal action [can influence] business activity and [shift] the sense of what represents a normal or desirable way of life. Growing enthusiasm for plant-based diets [which significantly reduce greenhouse gas emissions] prompted producers to invest in developing new vegan and vegetarian products" (Rogers, 2021).

WHAT YOU CAN DO

What Can I Do to Slow Global Warming?

Many ideas will be discussed in this section. We'll start with one that is admittedly touchy.

Choosing to Not Have Children (Or as Many)

No one should be forced to have children or guilted for not having them. It would be naïve, however, to think that climate change won't play a part in at least some young people's choice to have children or not, or how many.

The rising specter of droughts, heat waves, fires and floods from the climate crisis has put a level of despair in the air. A poll by the *Washington Post* and the Kaiser Family Foundation in September 2019 found that sixty-eight percent of respondents ages 18 to 29 say they are "afraid" of the effects of climate change. Sixty-three percent of teen respondents believe future generations will be harmed a great deal (Belkin, 2019).

Many of us made our choice about having children or not long ago. But for women of childbearing age, as one group's website puts it, the climate crisis is a reproductive justice crisis.

The group *Conceivable Future* (https://conceivablefuture.org) was founded in 2014 by two young women, Josephine Feroretti and Meghan Kallman. The group's website indicates it is "a woman-led network of Americans bringing awareness to the threat climate change poses to reproductive justice, and demanding an end to U.S. fossil fuel subsidies."

Neither woman, both in their thirties, had yet chosen whether to have children or not. They emphasize the group doesn't push women to choose one way or the other—that's an individual, personal choice. Two issues come up for women the most, though, says Kallman: "What harm will come to my child?" from living during the climate crisis and, "What harm will my child cause the world" due to climate change?

Emma Lim, who was eighteen at the time of a 2019 article regarding *Conceivable Future*, is one example. A pre-med student at McGill University at the time, she had "'grown up with the shadow of climate change looming over her,'" but "'when you are young you have a lot of faith and trust in your government. Mine has eroded over time.'"

While she was protesting weekly in front of the provincial legislature in Ontario, Canada for stronger actions on climate change, "a younger protester standing next to her, a girl in sixth or seventh grade, "'was holding my hand and crying. I really wanted to tell her it was going to be OK, but I couldn't,'" Lim remembered in an interview with Yahoo News. "'I just kept telling her I was sorry.'"

"She continued: 'I decided then that I couldn't have children. As a parent you're supposed to be there for your children, protect them from the world. But there are some things you can't protect them from. I would rather have the pain of not being a mother than have the pain of being one who can't protect them.'"

"That was in September [2019]. When Lim arrived back home, she built a website with a friend and created a pledge: 'I pledge not to have children until I am sure my government will ensure a safe future for them.' It now [in 2019] has more than 5,000 signatures from across Canada, the U.S., Europe, South America and Australia'" (Belkin, 2019).

Kallman said, "the very decision process, not a specific outcome, is what sets this generation apart. It is their version of the Depression, the Cold War, the overpopulation movement that gave some among their parents and grandparents reason for pause. It also clarifies the risks of climate change in personal ways" (Belkin, 2019).

Camila Thorndike, 32, said she was still "mired in ambivalence." The answer to this question, she said, will finally tip her decision one way or the other: "'How would I answer my child in 10 or 20 years' time, when they ask me, 'Why am I here? Did you know this was going to happen? What did you do to stop it?'" (Belkin, 2019).

To pursue the question of having children or not from the perspective of a local young male, I asked Ryan Hanna about it in our June 2022 interview. "Yes," he said he'd thought about having children in a climate change-driven future: "I think I'd like to have kids when I'm ready," but not now, with "really high" housing and food costs. He identified the cost of living, population growth and climate change to be the deciding factors on when he has children.

Fewer people than ever in our country are bringing children into the world. According to the U.S. Census: "The U.S. population grew at a slower rate in 2021 than any other year since the founding of the nation" (Rogers, 2021). Historically high inflation, energy, grocery and housing prices, plus the pandemic all have contributed to the plummeting birth rate, the article noted.

Finally, a friend noted the value of having a smaller family. Choosing to have two children instead of four, he said, lowers the tax burden of public schools, allows more time, energy and attention as well as raises the quality of life for all, allows for a better education and a closer relationship with parents.

According to a 2020 study, the top ten individual actions to slow global warming are (in order of descending effectiveness):

1. Live car-free
2. Shift to a battery electric vehicle
3. One less flight per year (long-distance return)
4. Purchase/use renewable electricity

5. Shift to public transport
6. Vegan diet
7. Shift to active transport
8. No pets
9. Heat pump
10. Renewable-based heating

(Source: Ivanova, et al., 2020, Figure 7)
https://iopscience.iop.org/article/10.1088/1748-9326/ab8589#erlab8589s4

Notice that the top three individual actions (and five of the top seven) involve transportation — reducing car and air travel, plus shifting to public and active transport. Three actions are housing-related: purchasing and using renewable electricity and renewable-based heating, as well as using a heat pump. The final two actions involve eating a vegan diet and having no pets.

Since transportation is the biggest emitter of carbon dioxide in Idaho (fifty-eight percent) (Hall, 2019), let's begin individual actions there. But first, a review of our auto-centric culture.

The Incessant Need to Drive

"Our location's greatest impact on our carbon footprint comes from how much it makes us drive."

Walkable City: How Downtown Can Save America, One Step at a Time, 57
author, Jeff Speck

"As long as we design our streets only for cars, we are designing a high-carbon future."

Emiko Atherton,
Washington Post author

WHAT YOU CAN DO

Since the end of World War II, communities have gone from being walkable to auto-centric. In so many cities and towns, a car is relied on to get to the grocery store, dining establishments, entertainment, parks and schools. Vehicles are served, not people.

Due to what's called a location's Walk Score (https://walkscore.com), it's easy to measure how walkable services are in a community, compared to the 1960s. Walk Score is a private company, launched in July 2007, that gives a walkability score for any address in the U.S. The lower the score the less walkable the location. The company provides a bike score for many locations as well (the lower the score the less bikeable the location).

Let's see how walkable and bikeable my location in Coeur d'Alene is: Walk Score: **69 out of 100** (some errands can be accomplished on foot). Not bad. My home's bike score? Even better: **72** (very bikeable; biking is convenient for most trips). I'm fortunate since Coeur d'Alene's average walk score is only **35** (most errands require a car). Post Falls is even worse — **23** (almost all errands require a car).

When most services are accessible only by vehicle, both climate-wrecking carbon dioxide and tailpipe pollutants (such as ground-level ozone, nitrous oxide and particulate matter) are produced. These are unhealthy to breathe, especially for the elderly, children and those with a heart condition, COPD, allergies and asthma.

With more cars on the road, traffic congestion gets worse. Also, there's more wear and tear on the roads and households have to spend more money to get around. More noise is produced as well and there are fewer social interactions in the community.

A former city planner in Vancouver, British Columbia and Calgary, Alberta, Brent Toderian, expressed it well to a group in Spokane in the summer of 2018: "When you design a city for cars, it fails everyone, *including drivers.*"

When someone at his public lectures invariably says, 'Don't you know I absolutely need my car?' Toderian replies, "**If you need your car, it's better for you if he rides his bike, and she takes transit, and he walks. Because if you're all fighting for the same amount of space, your car can't move very far. That's geometry**" (Deshais, 2018).

In discussing how to reduce transportation's greenhouse gas emissions, the prominent urban thinker and consultant said the idea is to design a multimodal city: "When you design a multimodal city that makes walking, biking and transit enjoyable, it works better for everyone, including drivers" (Deshais, 2018).

Expand Bicycle Use

In that same 2020 study of individual consumption changes needed to mitigate climate disruption, a shift to active transport (primarily bicycling) is number seven. Bicycle use also supports living car-free.

A basic way to understand the connection between CO_2 and transportation: each gallon of gasoline, when burned in an internal combustion engine, emits about twenty pounds of carbon (NASA). Therefore, the world needs to shift to battery electric vehicles, public transport and hybrid electric vehicles as quickly as possible.

I began riding a bicycle at age seven. Now at seventy-two, I continue to ride whenever I can, year-round. It's my favorite form of in-town transport unless it's raining, too cold or too risky on the ice.

For me, it's invigorating to ride around town. I arrive more energetic and focused than if I'd driven. The roads, with bike lanes, are safe enough to ride to visit friends, enjoy the parks and the lake, as well as run errands. I also appreciate how agile riding is. I can cut across parking lots and out of car lanes to get to where I want to go in ways I can't in a car. It gets me out of the house, too.

Bicycling also keeps me nimble, in both brain and body. I appreciate being in the moment to react to obstacles on the road (e.g., puddles, potholes, sewer grates) while navigating the quickest, safest way to get where I'm going. I enjoy the chill of an early January morning ride to the dentist's office and back. My calves come alive, my breath deepens, as I push up short hills. None of this sensory stimulation, and connection with the idiosyncrasies of home, are available driving a car.

Riding also helps my waistline, pocketbook and blood pressure. I sleep like a baby, too. Any muscle soreness early in the season is well-earned and short-lived. More people today are using e-bikes. They allow those who've been injured or have medical restrictions to go places they couldn't on a traditional bike, which can only be a good thing. However, I appreciate the exercise and self-reliance of pedaling my traditional bike.

Although I've felt the rewards of bike riding for years, I looked more closely into its health benefits; they're substantial.

A UCLA study found that riding a bike:

- Strengthens your immune system and helps decrease inflammation.
- Stimulates and improves your heart, lungs and circulation.
- Builds muscles in the upper and lower body.
- Improves mental well-being. Bike riders get a hit of endorphins, the brain's natural pain relievers, and triggers a positive feeling.
- Improves brain functions like memory and creative thinking.
- Decreases levels of stress hormones like adrenaline and cortisol, so you feel more relaxed.

- Boosts your energy; increases your stamina and endurance.
- Helps you sleep better.

Also, since more than half of all daily trips in the U.S. are less than three miles long, that's a perfect distance for a climate-friendly bike ride (UCLA, 2022). This brings up its substantial climate benefits. Biking once a day, research from the University of Oxford indicates, decreases the average person's carbon emissions from transportation by 67% (Hallisey, 2021).

Separate research showed the carbon footprint of cycling is "up to 30 times lower than that of a fossil fuel car, and even less than that of walking or taking public transportation" (Brand, et al., 2021). Also, the carbon footprint of bicycling is about ten times lower than driving an electric vehicle (Brand, et al., 2021).

Finally, bicycling can contribute more quickly to tackling the climate crisis than electric vehicles. How? Since it will take fifteen to twenty years to turn over all of the world's internal combustion engine vehicles to electric ones (*Future*, 2022). We don't have that much time to halve our greenhouse gas emissions by 2030.

Bicycling, then, is a simple way to reduce our carbon footprint. It's easy, healthy, gets us outside, saves money and reduces traffic congestion. Either by two wheels, or just walking, active transport by more of us can help restore our home.

In Coeur d'Alene we're fortunate to have a robust bike lane system and miles of dedicated bike trails. That's due to the efforts of people like Monte McCully, City Trails Coordinator, and Chris Bosley, City Engineer, as well as the city's Pedestrian and Bike Advisory Committee.

I interviewed Bosley in February 2022. He's not only a state and national award-winning engineer, but for many years a dedicated bicyclist. Every day, for fifteen years, he rode to and from work in Boston and here in town.

In 2019 Bosley championed, along with Russ Hersrud, co-founder of a local climate change group with the author, unanimous passage of the Complete Streets Ordinance by the Coeur d'Alene City Council. The law requires equal and safe access by bicyclists and pedestrians to all new city public street projects. This includes sidewalks on both sides of the street, extending existing bicycle lanes, flashing beacons for safe pedestrian crossings, as well as good pedestrian signage and covered bus stops.

Bosley described the Seltice Way project as the city's best example of a Complete Street and environmentally-friendly project. An increasingly congested two-way road, connecting Coeur d'Alene with Post Falls, Seltice Way became a well-designed flow of traffic.

This roadway now has buffered bike lanes, a shared use path (for bicyclists and pedestrians), roundabouts, as well as rapid flashing beacons for pedestrians to cross safely at uncontrolled, marked intersections. Bosley proudly noted that the bus stop shelters were built by the contractor from trees that needed to be cut down for the project. In addition, concrete from the former U.S. Hwy. 10 was crushed and used as fill at the site.

Since the interview, a bicycle boulevard on 5th St., from downtown to Harrison Ave., has been added. Kathleen Avenue, from Government Way to 4th Street, has had sidewalk gaps filled in and a second bicycle lane installed.

What about bicycling injuries in town? I asked. Most are from bicyclists riding the wrong way, while a car is backing out of a driveway, Bosley said. After a sixty-two-year-old woman was killed attempting to cross Best Avenue on foot at night, the city sought, and received, a grant to put in flashing lights at a crosswalk, plus a continuous sidewalk from 4th to 7th Streets.

Nationwide, more than three-fourths of people surveyed said being within an easy walk of other places and amenities such as shops and parks is important when deciding where to live (National Association of REALTORS, 2023). It's true here in the Lake City as well. In a September 2015 poll, ninety-four percent of Coeur d'Alene residents said they wanted the bike lane system to be expanded (email from John Kelly, March 17, 2016).

Most of us want walkable and bikeable communities and downtowns that are vibrant, safe and welcoming. Designing communities with these qualities brings in businesses that want to invest.

Families want to stay and shop as well as mingle with other community members as they enjoy the outdoors. Children and teens can play and engage in more independent activities, away from social media. Greenhouse gas emissions decline while personal and community health improve.

What's next, then, to expand bicycling in town? First, have the city proclaim an annual Bike to Work Week in mid-May, supported by the City's Pedestrian and Bicycle Advisory Committee. That consistency would encourage more businesses to pledge financial support and add to the number of people participating in the community rides.

Second, expand the walkability (and therefore the bikeability) of the city, making use of other cities' successfully implemented measures. More on this idea shortly. Third, support the trial run of the Lake City's first protected bike lane, along Mullan Avenue, from 19th to Coeur d'Alene Lake Drive. The project will use plastic barriers (filled with sand or water) to separate bikes and auto traffic. Protected bike lanes provide riders a greater sense of safety. If successful, there would be added motivation to find additional sites, and expand biking infrastructure.

Walk More

Although not listed as an individual choice to reduce extreme climate change, going for a walk is a simple way to connect to our home, be physically active and minimize automobile use.

Being close enough to walk to neighborhood shops, parks, schools and other amenities saves on gas money, wear and tear on the car, minimizes tailpipe air pollution and reduces congestion. It also provides relaxing exercise, and for many people, time to connect outdoors with their pet.

Research shows that those of us who live in highly walkable neighborhoods are more likely to engage in adequate physical activity and are thinner, compared to those in lower walkability neighborhoods (Wang, et al., 2023). Car dependent neighborhoods too often result in inactive lifestyles and obesity, which are responsible for up to ninety-five percent of diabetes cases in the U.S. (WebMD, 2023).

In our country, more than 100,000 annual "premature deaths" are attributed to air pollution (Goodwin, et al.). Approximately twenty-seven million Americans — 8.3% of us — suffer from asthma (Most Recent Asthma Data, CDC). But asthma is responsible for only a fraction of premature deaths. A generation ago, the leading cause of those deaths was from factories. Today it's from vehicle emissions.

Walkability matters to home buyers as well. Homes within walking distance of schools, shopping, parks and other urban amenities sell for 23.5 percent, or $77,665 more than comparable properties that are car dependent (Katz, 2020).

Walking also allows us to see the little things that we miss driving by in a car. Fun things like book sharing boxes in neighbor's yards and a mailbox with dog biscuits inside (and "dogs laugh with their tails" written on the mailbox door). On a recent walk with my partner's 14-year-old granddaughter, Ilah, we took a biscuit for Mallee, a delightful Australian shepherd mix. We'd miss little treats like that without going for a walk (and Mallee would miss hers!).

Speaking of Ilah, walks are also a great way to connect with the younger generation. On walks around the neighborhood or to the library, she talks to dogs and stray cats. She notices bugs and smells that I don't. I love that Ilah calls me "Grandpa," even if I'm not really her grandfather. We enjoy the sparkling Christmas lights and fabulous Halloween displays together.

My heart softens, sharing in a young girl's world. Our walks keep me younger and makes my appreciation of home that much richer.

Make Walking Safer

Walking in the United States, and in Inland Northwest towns, is growing more dangerous. The nonprofit association, "Spotlight on Highway Safety," projects that 7,508 people were killed walking on U.S. streets in 2022, the most pedestrians since 1981, when 7,837 were killed. Pedestrians struck by a large SUV are twice as likely to die as those struck by a car (Davis, 2021).

Here in Idaho, pedestrian fatalities rose seventy percent from 2009 to 2018, according to the American Automobile Association. In Kootenai County there have been seventeen pedestrian deaths from 2016 to 2020. This is according to information provided by the Kootenai County Sheriff's Office at the request of the *Coeur d'Alene Press*, for a January 2021 article ("Dangerous ground," *Coeur d'Alene Press*, January 13, 2021).

Vehicle-pedestrian collisions don't go well for the walker. If a pedestrian is hit by a vehicle going forty miles per hour the fatality rate is eighty-five percent; at thirty mph it drops to forty-five percent; and at twenty mph it's only a five percent fatality rate (*Federal Register*, 2015).

For pedestrian and bicyclist safety, speed limits on wide, multi-lane roads are too high. Transportation for America also notes more people are driving trucks and SUVs, compared to twenty years ago. Those trucks and SUVs are significantly bigger than sedans ten to twenty years ago. Therefore, they're more dangerous for people walking, biking, or getting around with an assistive device. Electric vehicles, with batteries, are even heavier.

Since 2009, the share of all new vehicles bought and sold in the U.S. that were sedans (with lower front bumpers and higher all-around visibility) plunged from nearly forty percent to below thirty percent. If current trends hold, that could be below fifteen percent before the 2020s are over (*Federal Register*, 2015). Also, pickup trucks and SUVs are two to three times more likely than smaller personal vehicles to kill people walking (*Federal Register*, 2015).

What can be done about the deadly trend? Pedestrians need to wear bright clothing. Cities like Coeur d'Alene need to continue to extend sidewalks, as well as build roundabouts and flashing crosswalks. Also, take advantage of federal funding. During a February 2023 television interview, Secretary of Transportation Pete Buttigieg said funding from the Bipartisan Infrastructure Law was available for cities and states to reduce the carnage (PBS Newshour, 2023).

Spokane was a recent recipient of that federal funding. In November 2023, it was awarded a $9.2 million grant, along with two million dollars of its own money, to make its streets safer for pedestrians and bicyclists. Nearly four million dollars will be used for new and updated bike lanes. An additional $5.64 million will go to improve intersections with accessible signals, curb extensions, high-visibility crosswalk markings and ADA-accessible ramps (Sullender, *The Spokesman-Review*, 2023).

Surging, then Slowing, Electric Vehicle Sales

Shifting to battery electric vehicles and hybrid EVs are effective individual actions to mitigate climate change. Commitments by car companies has been nearly unanimous. Volvo's chief technology officer, Henrik Green, said it's phasing out production of all cars with internal combustion engines, noting: "There is no long-term future for cars with an internal combustion engine" (*The Spokesman-Review*, March 3, 2021).

General Motors and Volvo committed to being all-electric by 2030 (*The Spokesman-Review*, Jan. 29/ Feb. 16, 2021) and Jaguar by 2025 (Pylas, 2021). Ford vowed to convert its entire passenger vehicle lineup in Europe to electric by 2030. This would meet lower limits on carbon dioxide in the European Union, to comply with the 2015 Paris climate agreement (Krisher & McHugh, 2021).

GM plans to have a lineup of thirty all-electric models worldwide by the middle of the decade (Krisher, 2021). In August 2022, Honda said it planned to phase-out fossil-fuel vehicles by 2040 (*The Spokesman-Review*, Aug. 30, 2020).

Toyota hasn't announced when it will go fully electric, although speculation is by 2035 (Horn, 2022). In July 2023, Toyota claimed a breakthrough in EV battery technology. It said it would manufacture solid-state batteries with a range of 745 miles and would charge in ten minutes or less (Davies, 2023).

Solid-state batteries use solid ceramic material, instead of liquid electrolytes, to carry the electric current, allowing it to store more energy, charge faster and offer greater safety. "Far less graphite and cobalt" than lithium-ion technology would be needed (although up to thirty-five percent more lithium), causing thirty-nine percent fewer emissions during manufacturing (L., 2023).

Toyota says it will introduce solid-state batteries as soon as 2027, with half of their planned 3.5 million EVs sold annually by 2030 to be solid-state. First-generation solid-state packs would have about 520 miles of range, with a ten minute 10-to-80% charge capability. Eventually the peak range would be up to 630 miles (Duff, 2023). The EPA says that the planet-warming pollution electric vehicles create over their lifespan, including battery manufacturing, is significantly less than gas-powered vehicles (https://www.epa.gov/greenvehicles/electric-vehicle-myths).

California, with the fifth-largest economy in the world, said it will end fossil-fuel vehicle sales by 2035 (*The Spokesman-Review*, August 25, 2022). Shortly after, Washington state followed suit (Kroman, 2022).

By the end of 2024, Tesla announced it would make at least 7,500 of its charging stations across the U.S. available to all EV users and compatible with Detroit manufacturers like Ford and General Motors. Tesla also partnered with Hilton Worldwide to install 20,000 universal electrical vehicle chargers at 2,000 of its hotels in the U.S., Canada and Mexico.

Walmart got into the act as well, saying it will add electric vehicle charging stations to thousands of its U.S. stores by 2030. Fast-charging stations will be added to Walmart and Sam's Club stores coast-to-coast, more than quadrupling its current network of roughly 280 locations.

Walmart's senior vice president of energy transformation, Vishal Kapadia, added that since "We've got a Walmart store or Sam's Club within 10 miles of 90% of the population in this country, we know we can address range anxiety in a way that no one else can" (Whalen, 2023).

The U.S. charging network has vastly expanded in recent years. There are more than 63,000 places to plug in on the road, scattered across the U.S. and Canada (Laing, 2023). As with any transition technology, though, there are bugs to work out. Not all charging stations work, there aren't enough of them that any EV owner can use and they're unequally distributed.

Electric vehicle sales increased by fifty percent in the U.S. in 2022. By the end of 2023, about nine percent of cars sold in the U.S. were electric, up two points from 2022 (Gerdes, 2024). Sales had increased by forty-seven percent. However, in the first quarter of 2024, demand for EVs slowed considerably, to 2.7 percent (The Guardian, 2024), due to high prices and spotty infrastructure.

When that slowdown continued, Ford shifted its EV strategy. In mid-April, it delayed some EV models, cut spending on battery-powered vehicles by twelve billion dollars and pivoted to offer gas-electric hybrid propulsion systems across its lineup (Naughton, 2024). Tesla also cut ten percent of its workforce.

In the spring of 2023, the Environmental Protection Agency developed mileage rules so strict that the only way carmakers could comply would be to sell a tremendous number of EVs in a short time frame. By 2032, sixty-seven percent of sales of new cars and light-duty trucks would need to be all-electric, up from less than 7.6 percent in 2023.

Due to pressure from car manufacturers, Biden administration officials tweaked the plan in February to allow the sharp increase of EV sales to happen after 2030, not before (Davenport, 2024).

China's leading auto brand, BYD, sells the most EVs in the world, while Tesla continues to dominate the North American market. A

Tesla stockholder I know believes they'll be manufacturing twenty million EVs a year by 2030. We'll see, as Tesla faces increasing competition, including the introduction of much cheaper Chinese EVs in the U.S. in the coming years. By mid-century, with the power sector more than eighty percent zero-carbon, it's expected that more than three-quarters of all driving will be in EVs (Bullard, 2023).

Back home, Idaho is seeing modest gains in electric vehicle sales. In 2021, there were 2,990 registered. By 2023, Idaho had 5,940 electric vehicles registered (Nasdaq, 2024). A network of charging stations every fifty miles along the interstate will be built over five years with Idaho's $28 million share of Inflation Reduction Act funds (Corbin, 2022).

EV infrastructure in Coeur d'Alene is expanding as well. In mid-August 2019, an Internet search showed the Lake City had three electric vehicle charging stations. Less than two years later, in April 2021, forty-three electric charging stations were noted. In late August 2022, ninety-six, according to PlugShare (https://www.pluginamerica.org). Most are likely at-home units.

In addition to the substantial climate benefits to electric vehicles, foregoing gas-burning cars will have substantial health benefits as well. Princeton researchers found the transition to EVs could avoid as many as 165,000 deaths in this country by 2050. This is primarily due to reduced levels of harmful exhaust substances, such as particulate matter, nitrogen oxides and volatile organic compounds (Net-Zero America report).

Electric vehicles are also eighty to ninety percent efficient at converting energy to move a vehicle vs. seventeen to twenty percent efficiency for internal combustion engines, according to Stanford professor Mark Jacobson (Expert witness testimony, *Held v. State of Montana,* June 2023). The average EV owner will spend sixty percent less to power their vehicle and half as much on repairs and maintenance — no oil changes or tune-ups — as the owner of a gas-powered vehicle (Barry, 2023).

It will take time to develop the infrastructure for electric vehicles. After all, gas stations and roads were not available everywhere when internal combustion engine automobile use began to take off. The Rhodium Group's Trever Houser said as much, stating that the recent slow pace of EV sales growth is "normal and expected" (Brocious, 2024).

With additional quick-charging stations nationwide, and the federal tax credit of up to $7,500 on new electric cars (with certain exceptions; and up to $4,000 on used EVs) available until 2032, many of us will likely be purchasing electric vehicles in the future.

Flying Less

One less long-distance flight a year and one less medium-distance flight a year are recommended as personal changes to mitigate climate change. Again, transportation, in whatever form, is a key player in climate disruption, and restoring our home.

I haven't flown out of the country since 2019 but will in August 2024, to Norway. I have little knowledge of the impact of long- or medium-distance flying on greenhouse gas emissions.

Renewable Electricity

Number four in individual changes to minimize climate change is producing/using renewable electricity; using renewable-based heating is number eleven.

I've never used solar panels or a wind turbine to generate my own electricity. With prices of solar panels dropping ninety-nine percent over the past four decades (Chandler, 2018), though, I've been curious about purchasing them. Fortunately, my partner Ginny's house has better south- and west-facing sun than my previous location. The next-door neighbor, it turns out, is a solar salesperson as well (a nice coincidence). She's looking into our house being a test case for local solar photovoltaic use.

Nationally, renewable energy hit a new milestone in the first half of 2023, generating twenty-five percent of all electricity. Wind contributed ten percent, hydropower six percent, solar three percent, biomass one percent, and geothermal less than half a percent (Energy Information Agency's "Power Monthly", June 2023). In Idaho, hydropower generates about fifty-three percent of our electricity, natural gas nearly twenty-five percent and wind just over seventeen percent (U.S. Energy Information Agency, 2022).

Use Heat Pumps

Heat pumps have hit the mainstream. In 2023, for the fourth year in a row, they outsold gas furnaces in America. Incredibly efficient, a heat pump can produce the same amount of heat as a fossil-fired system using "just a third or a quarter of the energy" (Takemura, 2024). That's important since heating a building's air and water by fossil fuels, primarily by natural gas around here, and cooling with an air conditioner, generates one-third of all greenhouse gas emissions.

A heat pump is basically a refrigerator in reverse. "Heat" from the outside air (down to minus ten degrees F) is absorbed by a colder refrigerant, which warms it until it becomes a gas; the gas is then compressed, creating more heat. The air is then blown over the heated coils and distributed throughout the house, warming it in the winter. A changeover valve reverses the process in the summer, to cool the house, even at 110°F.

Changing from a natural gas furnace to a heat-pump system is estimated to reduce emissions by sixty-two percent to sixty-eight percent in Spokane, according to a 2022 peer-reviewed study (Pistochini, et al., 2022). A tax credit of up to $3,200, including labor for installation, is available for heat pumps (https://www.irs.gov).

Consumer Reports says cold-weather heat pumps can work quite well in our winters, as long as they're properly sized and installed by an expert in heat pump systems. Equipped with a variable speed compressor that holds a steadier temperature than a traditional, single-speed HVAC system, it helps to save energy (McCabe, 2022).

Exciting news in 2023, too. A startup company, Gradient, has developed a heat pump that can be installed in a window, similar to an air conditioner. It doesn't require an electrician or plumber to install it. The unit straddles a window sill so the power and noise are outside the window, the heating and cooling inside. The greenhouse gas emissions reduction is up to ninety-five percent, depending on your source of electricity, according to the company website. "That's like choosing to walk instead of drive up to 15 miles daily," it adds.

The heat pump uses a next-generation refrigerant with a lower environmental impact. Pre-orders for large multi-family projects are being taken now, with a waiting list for individual home sales (https://www.gradientcomfort.com). The 120-volt units will hit the market about September, a January 2024 email noted. It's likely the $169 million in grants from the Department of Energy to nine companies, including Gradient, will boost domestic production and lower costs (Pontecorvo, 2023).

Avista Utility representatives have expressed concern about the greater draw on their electric grid from heat pump systems. It's expected that installing systems in new commercial construction and multifamily homes first, as now mandated in Washington state, will allow utilities to find more and cheaper clean energy sources in the future.

Finally, an aside, on the cooking scene: induction cooktops, by directly heating the pot, can boil water in half the time (Expert witness testimony, Mark Jacobson, *Held v. State of Montana*, June 2023). They also don't threaten home air quality, as natural gas stoves do, as was announced in June 2023 (*The Spokesman-Review*, June 22, 2023). As costs come down, these will be a more viable alternative as well.

Wash Clothes in Cold Water and Less Often

Water heating consumes about ninety percent of the energy it takes to operate a washing machine, according to Energy Star. A *Washington Post* article states that changing from washing in hot to warm water can cut your energy use in half. Washing with cold water will reduce your energy footprint even more. By washing four out of five loads of laundry in cold water, you could cut 864 pounds of CO_2 emissions in a year, equivalent to planting 0.37 acre of U.S. forest, according to the American Cleaning Institute (Chiu, 2024).

Try to run your machine only when it's full. And wash your clothes less often. "Every time we put that garment in the washing machines, part of it is gone down the drain," Cosette Joyner Martinez, associate professor in the department of design, housing and merchandising at Oklahoma State University, previously told the *Washington Post*. "Then we lose another piece of it in the dryer in the lint trap, so we're disintegrating our garments."

Two other ideas: to limit your exposure to plastic, use laundry powder rather than polyvinyl alcohol (PVA)-wrapped pods and sheets, the *Washington Post's* Michael J. Coren writes. Whenever possible, air dry your clothes. That saves the most energy and is better for your clothes. Put shirts on hangers to reduce wrinkles and lay flat heavier fabrics, such as knits (Chiu, 2024). Personally, I enjoyed getting the clothesline out in mid-April, to let the sun do the drying.

Expand Public Transportation

In Kootenai County, public transit is provided by Citylink North. A rural route system further south, in Benewah County, serves towns on the Coeur d'Alene Tribal Reservation. The service is free.

The Kootenai County system provides service to Coeur d'Alene, Dalton Gardens, Hayden and Post Falls, six days a week (6:00 am to 6:00 pm M-F and 9:00 am to 3:00 pm on Saturday; not Sundays). Federal money, as well as grant matching by participating cities and voluntary investments by Kootenai Health and the Coeur d'Alene Tribe fund the Citylink system. As many as fifty thousand people per month were transported, pre-pandemic, at over one hundred stops ("Public Transportation Plan Update," 2012).

In late November 2023, I spoke with Jody Breze, director of Citylink North, about expanding Citylink to Rathdrum. She explained that time and money are the key constraints to expansion. A full transit loop takes one hour. A private vehicle can make the same trip in less than thirty minutes. To match that would require the use of multiple buses at additional costs. Citylink continues to seek additional funding from employers and the educational community, to expand service to outlying areas. A shortage of bus drivers is also a constraint to expansion, Breze said.

Eat a More Plant-Based Diet

"Nothing will benefit human health and increase the chances of survival on earth as much as the evolution to a vegetarian diet."

Albert Einstein

Not eating meat is another important individual choice to mitigate climate disruption. Specifically, eating a vegan diet (no meat or dairy products) is number six on the list; being a vegetarian (with dairy and eggs, but no meat) is number thirteen.

I chose to become a vegetarian in 1988, well before I was aware of global warming. While being a vegetarian, let alone a vegan, is not for everyone, it's not been difficult for me. Here's how it happened.

I began working for a developmental disability company in the late 1980s in Post Falls. During a conversation with a fellow employee, Mark asked: "How can you be an environmentalist and eat meat?" The question caught me by surprise; I'd never thought there was a conflict between the two.

During my next days off, the question weighed on me. After much thought, when I returned to work, I told Mark I'd made the commitment to stop eating meat. I've been meat-free since, for the past thirty-six years.

I'm an ovo-lacto vegetarian, one who can eat eggs and drink milk. I have eggs in some form a few times a month, and I haven't drunk milk in years. Initially, I gave up eating meat because I'd become tired when I did so, similar to the tryptophan-induced drowsiness after eating turkey at Thanksgiving.

Later, the health benefits became important to me. In addition, I oppose the cruelty and slaughter of animals. Being vegetarian is also consistent with my value of "walking softly on the earth," a

Native American saying. Finally, it supports being frugal; not eating meat has saved me a lot of money.

Consider:

> The Western diet comes with a steep climate price tag. The most conservative estimates suggest that raising livestock accounts for nearly 15 percent of global greenhouse gases emitted each year; the most comprehensive assessments of direct and indirect emissions say more than 50 percent.... The production of meat and dairy contributes many more emissions than growing their sprouted counterparts—vegetables, fruits, grains, and legumes. Ruminants such as cows are the most prolific offenders, generating the potent greenhouse gas methane as they digest their food. If cattle were their own nation, they would be the world's third-largest emitter of greenhouse gases. Overconsumption of animal protein also comes at a steep cost to human health.... On average, adults require 50 grams of protein each day.... In the United States and Canada, the average adult consumes more than 90 grams of protein per day. Where plant-based protein is abundant, human beings do not need animal protein for its nutrients (aside from vitamin B12 in strict vegan diets), and eating too much of it can lead to certain cancers, strokes, and heart disease.... (Drawdown, 2016, p. 39)

"According to the World Health Organization, only 10 to 15 percent of one's daily calories needs to come from protein, and a diet primarily of plants can easily meet that threshold" (Drawdown, 2016, p. 39).

A groundbreaking 2016 study from the University of Oxford modeled the climate, health, and economic benefits of a worldwide transition to plant-based diets between now and 2050. Business-as-usual emissions could be reduced by as much as seventy percent through adopting a vegan diet and sixty-three percent for a vegetarian diet (which includes cheese, milk, and eggs). The model also calculates a reduction in global mortality of six to ten percent (Springmann, et al., 2016).

Many people are not willing to give up eating meat, though, including my two sons and nearly everyone else I know. People love the taste, texture and cooking of meat. A better alternative for many is plant-based meat substitutes, including those offered by Impossible Foods and Beyond Meat, among others. I enjoyed an Impossible Foods hamburger at a local restaurant on my sixty-ninth birthday. It looked, tasted and felt like the hamburger of yesteryear.

"Meatless Monday," not eating meat one day a week, is likely a more sustainable option for many people. Studies show that small changes are more likely to last than significant lifestyle changes. According to a University of Michigan study, even if only about two of every five Americans participated each week, that'd be like taking more than 1.6 million cars off the road each year (Heller, 2020).

Cut Food Waste

Recently, wasting less food has been emphasized in some local grocery stores and on television. In America, we waste up to forty percent of our food supply. When that wasted food rots, it produces enough methane to cause as much warming as 3.3 billion tons of carbon dioxide each year (Gates, 2021, 121). The U.N. estimates that if wasted food was a country, its greenhouse gas emissions would rank third globally, behind China and the U.S. (Gates, 121).

So how do you and I prevent food waste, in a way that can make a climate change difference?

One way is how I was raised.

My parents grew up during the Great Depression. We didn't get dessert if we didn't eat everything on our plates. Mom was smart about it too; in order to have chocolate cake for dessert we had to eat things we didn't like. It was her way, Mom said, to have us try new foods, not waste and appreciate what we were given. It wasn't a big hit with my sister, though. Janet said she gained too much weight because of it.

The idea may not be popular in some families also. It's best to have a meal plan and buy food in quantities you'll eat, to minimize waste and save money. Also, consider the cost of "cheap" fast food, which is not only unhealthy but produces a tremendous amount of packaging waste.

In the past year, the group Enviro Certified has taken on reducing food waste through its Food Rescue Recognition Program. Locally, Fred Meyer, as part of Kroger's Zero Hunger, Zero Waste program, offers customers left-over food items at much reduced prices. Safeway, Super One Foods and Grocery Outlet, here in Coeur d'Alene, are Enviro Certified as well.

In addition to individual and corporate initiatives, the U.S. and several states and localities have signed on to a global target of halving wasted food by 2030. Policies restricting food from going to landfills (called food waste bans) can help, leading to increased food recycling through composting or anaerobic digestion, and related infrastructure.

But the best way to prevent food waste on a large scale? Avoid unnecessary food production. Waste-ban policies should promote waste prevention and donation. California is requiring, by 2025, at least twenty percent of edible food that would otherwise be disposed instead be recovered to eat (*Post* Opinions Staff, 2019).

WHAT YOU CAN DO

Plant More Trees

As noted in the Forests & Climate Change section, trees provide many benefits, including improving air and water quality and reducing heating and cooling costs. Trees can cool cities by up to ten degrees (Twitter, March 7, 2023).

Arbor Day (Latin for "tree") is a holiday that celebrates the planting, upkeep and preservation of trees. In Idaho, Arbor Day is celebrated the last Friday in April each year. Free seedlings to plant are available from the Idaho Forest Products Commission (https://www.idahoforests.org/content-item/tree-seedlings/). Post Falls also has an annual tree giveaway in mid-April.

If you want to expand your impact further, join the non-profit Arbor Day Foundation. It has a goal to plant 500 million trees worldwide by 2027. It's already planted more than 350 million.

Planting more trees in poorer parts of towns, with more paved surfaces and higher summer temperatures, is an effective way to reduce extreme heat risks and income inequality. Maintenance and watering need to be part of that tree planting program as well.

Save Water

We are blessed here, in the greater Coeur d'Alene/Spokane area, to have the remarkably clean and large Spokane Valley/Rathdrum Prairie Aquifer. The aquifer provides drinking and irrigation water for more than 500,000 people, and interacts with the Spokane River for recreation, flood control and power generation for cities in the area.

The aquifer, however, is not endless. Its source, the melting of snowpack, is diminishing and melting sooner, due to climate disruption. With our area's booming population growth, the demand on the aquifer is increasing. It behooves us to learn from others how

193

to best conserve precious water. Doing so will save the money and energy to pump and purify its use as well.

The best example of how to conserve water comes from an unlikely source, the desert in Las Vegas, Nevada. The city of 2.4 million, plus its suburbs and forty million annual visitors, gets ninety percent of its water from the Colorado River, ten percent from its groundwater.

It's "taken the most dramatic steps to reduce its dependence on Colorado River water," said Anne Castle, a senior fellow at the Getches-Wilkinson Center for Natural Resources, Energy and the Environment (Naishadham, 2022). Here's how cities in the west, including Las Vegas, have water amid drought.

In 2018, I visited Las Vegas. None of what had been done to save water was evident to me. Sin City's fountains, swimming pools, and showers use recycled water. Once used, much of that wastewater is treated and then returned to Lake Mead, the reservoir behind Hoover Dam, before it is drawn and used again.

> Las Vegas started conserving, reusing and recycling water in 1999. Since 2002, the Southern Nevada Water Authority has slashed its use of Colorado River water by 26% while the region's population grew by 49%. In 2003, the water authority banned front yard lawns in new subdivisions. Grass was prohibited in new commercial developments. Last year, Nevada outlawed what it called 'non-functional turf' in the Las Vegas area, or grass used at office parks, in street meridians and at entrances to housing developments. Officials said the measure could save an amount equal to 10% of its Colorado River allocation (Naishadham, 2020).

Ultimately, though, climate change may limit what cities can do inexpensively, says Dr. Daniel Swain, a climate scientist at the University of California, Los Angeles. Water desalination is quite expensive.

"'There's an assumption baked into almost all of these drought mitigation strategies and plans and water allocations that in the long run, drought is temporary,' said Swain of UCLA.

'Increasingly, it's an assumption that is wrong.'"

"'Swain added that conservation is easier in its earlier stages. You fix leaks, put in (efficient) toilets and fixtures and things like that in urban areas. After a certain point, you then have to start going for the higher hanging fruit'" (Naishadam, 2020).

What can Coeur d'Alene learn from Las Vegas, to conserve water? Begin with the basics: limit the watering of lawns to certain days and times. Don't irrigate lawns between noon and six pm., for example. People with even-numbered addresses water on Monday, Wednesday and Friday; those living at odd-numbered addresses, on Tuesday, Thursday and Saturday. In fact, plants grow better root systems if they're not watered every day.

Coeur d'Alene needs to create their own water-saving plan. In addition, the Lake City should begin looking into recycling gray-water and homeowners to focus on landscaping using native plants. Water saved today will be critically available for climate induced, extended droughts of the future.

Minimize Air Conditioner Use

Air conditioners use man-made chemicals called hydrofluorocarbons (HFCs) as refrigerants. HFCs are thousands of times more potent at trapping heat than CO_2. Thankfully, the Montreal Protocol required countries to reduce the use of such chemicals, starting in Jan. 2020.

If you need to use an air conditioner, buy super-efficient models and encourage legislators to reduce HFC production by forty percent by 2030 (*Post* Opinions Staff, 2019).

In my previous location, I didn't own an air conditioner, to keep from adding to the climate change problem. Instead, I upgraded windows in my previous residence to triple pane, in May 2020. It worked so well I didn't need to use a fan to stay cool that summer. I did during the oppressively hot heat wave of 2021, though. And I learned a lesson which might be helpful to others.

I installed indoor shades on the west and south-facing windows, and over the sun roof in 2022. Those actions helped to keep the high August 90°F heat out. Sleeping overnight was a struggle at times, though, with no cooling winds, only the use of a fan. The previous modifications were helpful as well during the summer of 2023.

My partner's house hadn't needed air conditioning since it was built on a slab. But for our comfort in wildfire smoke and Airbnb customers' comfort upstairs, we had installed an efficient system, at a very good cost, in May 2024.

A Walkable City

A Walkable City has everything you want within about a twenty-minute walk. Walkable communities are generally compact, with good walking surfaces. They have direct, obvious and safe routes, with connections to places people need and want to go. These include grocery and retail shopping, schools, parks, medical or other services and social activities, including transit.

What are some of its benefits? They allow residents to drive less, reducing transportation expenditures, which average sixteen percent of people's income. Walking also reduces traffic congestion, air and noise pollution. Wear and tear on city roads, consumption

WHAT YOU CAN DO

of petroleum and need for additional roads and parking drop. In addition, pedestrian and motor vehicle-related crashes, injuries, and fatalities are minimized.

Walkable communities increase housing values, attract new economy workers as well as tourists. Business relocation opportunities and infrastructure investments increase. Commuting costs decrease. In addition, the level of physical fitness and general health improves as people walk or bike more, and social interactions in the community increase.

How does a walkable city work? Neighborhoods are analyzed for these three factors:

- Distances —how easy is it to travel by foot or bike.
- Destinations—the presence of nearby businesses (grocery stores, restaurants, and retail) and public facilities (schools, parks).
- Density—having sufficient numbers of residents, employees and income to support businesses and public facilities.

Distance

Portland, Oregon developed a walkability plan in 2014. It's now called Complete Neighborhoods. Emphasizing short distances for walking, the Portland Plan used an analysis area with dimensions of 500 by 500 feet. The frequency of intersections and the presence of sidewalks were factors in assessing walkability of neighborhoods. Slopes over twenty percent were seen to limit walking and biking accessibility. Transit availability, providing access to more distant destinations, was also a factor.

Destinations

Refers to the quality and type of the destination (e.g., the presence of nearby grocery stores, restaurants, and retail). Specific destinations to evaluate include: full-service grocery stores, chain and single-store operators, neighborhood-serving retail, eating and drinking establishments, parks, and elementary schools.

Density

Density is needed to support the services used as walkable destinations. Neighborhoods generally require higher residential densities than typically found where the car is the dominant mode of travel. In Portland, twelve to eighteen households per acre was the minimum density needed to support the retail uses selected as destinations.

In Coeur d'Alene, an analysis of the town could be performed first, to see what areas have a significant presence of all three neighborhood factors. Areas that have some elements and those which lack significant walkable neighborhood characteristics would be assessed.

If research here shows that the idea warrants promise, the city could support its implementation. This would require developing a goal of what percentage of population would live in such neighborhoods, by a specific date. Portland's Climate Action Plan, for example, suggests a target of ninety percent of Portlanders living in walkable neighborhoods by 2035.

Where to Invest Your Money

"The climate emergency is urgent and climate risk is financial risk."
Kristina Wyatt, oversaw SEC's development
of climate risk disclosure rules

For those with money to invest, making prudent choices to minimize climate disruption is important. Thanks to my parents, I have a few investments, along with my brother and sister. Ed, my financial advisor, has kept my investments, as much as possible, out of the fossil fuel industry.

During our annual visit, in February 2023, the conversation with Ed turned to ESG investing (focused on reducing environmental, social and governance risks). Historically, Ed said, ESG-focused companies have lower rates of return than non-ESG companies. Some companies use the ESG designation to "greenwash" their environmental impact as well. After reviewing the investment performance of Hartford Climate Opportunities Fund, Ed said he would add it to my portfolio.

Conservative politicians, including some in Idaho, label ESG investing "woke capitalism." House Bill 191 prohibits a public entity in Idaho from accepting or denying bids for contracts over $100,000 based on ESG factors. It was signed into law on March 23, 2023. Both of my state representatives and senator, I disgustingly discovered, voted for the bill.

I didn't sleep well after reading that news and hearing from Ed that the returns were not as good on ESG investments. I needed to do my own research, I determined.

ESG investing, a *Forbes* story explained, allows an investor to know what environmental and social risks their clients face by investing in a company's stock. Consider, for example, the $742.1 billion in U.S. losses the past five years due to extreme weather events (according to NOAA, the National Oceanic and Atmospheric Administration). That's more than one-third of the total disaster cost of the past forty-two years ($2.155 trillion) (Michelson, 2023).

If you'd invested in any of the homes or businesses that were wiped out or severely damaged by those catastrophes, you'd have lost *a lot of money*, the article said. Wouldn't you want to know if your investment is at risk? Heck yes, I answered; we all would.

That's what ESG investing does, the article said. But does it make a decent return? Yes, it makes a decent return in just over half the cases (fifty-eight percent of the time), according to a July 2023 report by SS&C, an investing technologies company. The return was neutral or had mixed results thirty-four percent of the time; only eight percent was negative.

Due to the devastating physical and financial impacts of climate-related events, the Securities and Exchange Commission has been developing climate risk disclosure rules. Kristina Wyatt, who oversaw the SEC task force that developed those rules, noted: "investors were saying, 'look, the information that we're getting just doesn't cut it. It's not sufficiently clear, consistent, comparable, reliable. We need…the SEC to…provide clearer guidance'" (Michelson, 2023). As noted earlier, the less extensive than hoped for rules were finalized in February 2024.

A portfolio manager at a New York City-based investment company, interviewed for a PBS Newshour story, explained the upside of ESG investing. Dan Abbasi of Douglass Winthrop Advisors, LLC indicated that there are "growth opportunities for companies selling solutions to climate change." Automobile companies shifting to electric vehicles and charging infrastructure companies come to mind. A good friend, in his early 40s, is living off his Tesla stock, for example.

Abbasi noted his company's strategy has "outperformed benchmarks like the S&P 500 in the MSCI World for 6.5 years." I didn't understand what that meant, even after I looked it up. Thankfully, Abbasi simplified the message: "It's just a prudent investment agenda that's very clear-eyed about the way the world is changing."

Rebecca Henderson of the Harvard Business School concurred: "Not to pay attention to what's happening to the climate, not to pay attention to the consequences of inequality of the society we're in,

WHAT YOU CAN DO

not to be aware of the new products and opportunities that dealing with these issues is going to create is to be blinkered, and shortsighted" (PBS Newshour, August 29, 2023).

Based on my research, I wholeheartedly agree.

One final point for those who invest their money. According to a February 2024 study, the energy sector (heavily weighted towards fossil fuels) has underperformed for most of the past ten years: "In eight of the ten years between 2012 and 2021, ... [it] trailed the performance of the Standard & Poor's 500, and in five of the years, it placed dead last" (Novel Investor: Annual S & P Sector Returns).

Even after the world emerged from the COVID-19 pandemic and the Russian invasion of Ukraine, and oil prices surged, resulting in record profits, "...the fossil fuel sector posted a -4.8% return in 2023 (Yardeni Research. S & P 500 Sectors & Industries. December 30, 2023).

The study's authors concluded: "The traditional thesis underlying the industry — that the fossil fuel industry and economic growth are inextricably linked — is eroding. Facing increased competition between fossil fuel producers and from cheaper alternative technologies, the industry is ill-prepared to manage shareholder value in the coming years" (Chung & Cohn, 2024).

The take-home message: it's not just ExxonMobil that's had poor investment returns. The entire fossil fuel industry is not a place to invest your money long term.

Community Adaptation Plans

While not offering a solution per se, adaptation plans inform a community what climate impacts are coming and how to prepare for them. Climate science is used to project what impacts a community will likely face in the future, based on different emissions scenarios. Those impacts are then used to recommend adaptations a community can take to protect its people, natural resources, and its plant and animal life.

To project the future climate is vital, since assuming the future will be like the past is foolish. To do so is to ignore climate change. Temperatures, levels of precipitation and other variables are projected to the year 2100. Based on both high emissions and moderate emissions scenarios, members of the adaptation plan team make recommendations on how the community can adapt to expected impacts.

For example, adaptation plans can help protect a community's water supply. They can protect the health of those living and working in heat waves, wildfire smoke and higher pollen counts. Such plans protect children, the elderly, as well as those who struggle with allergies, asthma, COPD and heart conditions.

Ski resort owners can learn how to adapt to declining snowpacks. Whitewater rafting companies and fishermen will need to adapt to more "skinny water". Also, fishermen will have to deal with more invasive, warmer-water fish in our streams and lakes.

In 2015, the Federal Emergency Management Agency (FEMA) began requiring states to consider the effects of climate change in their State Hazard Mitigation Plans. State's plans need to be approved by FEMA in order for them to receive federal funding for pre-disaster mitigation projects, including floodplain mapping. Instead of just paying for post-disaster recovery, with adequate Congressional funding, the idea is to avoid those losses by building community climate resilience.

Spokane and Coeur d'Alene Adaptation Plans

Under the guidance of the National Oceanic and Atmospheric Administration's-funded Climate Impacts Research Consortium (CIRC), a group from Spokane met in May 2018. They eventually became the Spokane Community Adaptation Project (SCAP). The group, composed of community members, researched and prepared the community adaptation report. They used the Northwest Climate Toolbox, CIRC's suite of online climate science tools.

Over the following year, they developed a report on the expected outcomes until the year 2100. It included increases in Spokane area temperature, loss of snowfall, reductions in stream flows, risks of wildfires and impacts on agricultural production. The group also developed resiliency actions to adapt to each expected outcome.

SCAP found, for example, that in the forty years since the 7.6-mile-long Bloomsday run had begun, increasing temperatures had caused "an increase in heat-related health issues." Focusing on the May 2018 race, "warmer-than-normal temperatures likely led to an increase in the dropout rate for the race." With temperatures that could climb over eight degrees F. by the end of the century, the group expects more heat-related health issues in the future.

The group found snowfall at area ski resorts will be further impacted by climate change. With winter temperatures by mid-century expected to be 4.7°-5.9°F warmer at the Mt. Spokane ski resort than the historical average, forty-two to fifty-eight fewer freezing days per year are expected. This will shrink the length of the ski season and lead to economic impacts.

Specifically, *by the end of the century, under a high emissions scenario, the adaptation project found skiing or snowboarding may be **impossible** at all five area ski resorts* — Mt. Spokane, Schweitzer, Silver Mountain, Lookout Pass and 49 Degrees North — at least on south-facing hills. It will simply be too warm for snow to exist, whether artificial or natural.

In January 2020, Coeur d' Alene embarked on its own adaptation project, coordinated initially by this author. For the past two years

David Muise has been its coordinator. Local residents, myself included, identified the following climate change-related concerns: the health of Lake Coeur d'Alene, wildfire danger, and the health impacts of rising heat and wildfire smoke. A summary of the report will be presented at the North Idaho Green Summit on May 18, 2024.

In addition to preparing a community for inevitable climate impacts, adaptation plans help to answer an important question: why should I care about global warming? Adaptation plans serve to motivate and educate the public. These plans help to reduce the community's risk to negative impacts from climate change and adapt to those already in the works.

After release of the Coeur d'Alene group's report, presentations will take the message to fellow residents, stakeholder groups, interested parties and policy makers. The goal is to both generate important conversations and present the report's recommendations, with structured follow through steps.

Climate-Aware Therapy

Lastly, a much different "what you can do," to conclude this section: see a climate-aware therapist. I realize this is a bit unusual but it seems interesting as well.

As an Idaho certified alcohol/drug counselor and a Masters level therapist for twenty years, I hadn't heard of a "climate-aware" therapist until I ran across the Climate Psychology Alliance of North America's website. A U.S. map on the site showed the names and locations of 192 climate-aware therapists.

Curious, I looked for the one closest to Coeur d'Alene. Much to my surprise, it was a PhD psychologist in Ellensburg, Washington, where my sister still lives, and I resided from junior high through undergraduate school.

WHAT YOU CAN DO

Curious, I called Marcia Rorty, PhD on September 7, 2022. Open and generous with her time, we spoke for twenty-five minutes. Dr. Rorty said she's been in Ellensburg for four years, having moved from Los Angeles, where she received her PhD.

How many clients would the doctor estimate she provides therapy for, with climate change-related issues, I asked? I was quite surprised: seventy-five to eighty percent, she said. Many don't initially present with a climate-related issue, but "it's a general issue in the backdrop…It just comes up," she added.

With older Medicare clients, for example, who were "full of so much optimism" in the '60s, now they're "petrified of going out on such a scary note," with heat waves and wildfire smoke.

They're especially "terrified" for their children and grandchildren. Dr. Rorty noted some are worried that their grandchildren will live in poor housing in a flood zone, at high risk of being swept away in a flood.

A teenager the day before who was "already hurting" when she arrived, shared that she was "frantic about climate change," feeling "everything was doomed." Another young client, who'd been looking to get into the ecology field earlier, had said to himself 'screw it, everything here is going to hell', so he decided to get a job where he "could make as much money as possible and smoke weed."

The day before, a Telehealth (by computer) client in Los Angeles, who was living in a basement with air conditioning while it was 108°F. outside, was "terrified for the homeless population there." The incredible heat and rampant, huge fires and smoke "so worry him."

What type of therapy does she provide for climate anxious clients, I asked? After attending a ten-week course by Portland, Oregon-based environmental psychologist Thomas Doherty, she emphasizes the importance of self-care: getting good quality sleep,

205

eating well and being out in nature. Doherty also has clients remember times they spent in nature as a child, making a map of those times in wild areas.

Dr. Doherty also does "walking therapy," where he walks and talks with clients on trails. Being in an enclosed room, sitting in designated chairs is "too intense for some clients."

A directory of climate-aware therapists is located on the Climate Psychology Alliance of North America website: https://www.climatepsychology.us/.

CHAPTER 13

What We Can Do Together

> "Hope is not the conviction that something will turn out well. It is the certainty that something is worth doing regardless of how it turns out."
> Vaclav Havel

> "The one thing we need more than hope is action. Once we start to act, hope is everywhere."
> Greta Thunberg,
> Swedish climate activist, author, creator,
> Fridays For Future movement

IN THIS CHAPTER, I emphasize actions to be taken in concert with others, to preserve our local gifts of nature. They're divided into local (city and county) actions, then national and international actions. We are, after all, dealing with *global* warming, albeit with local impacts. Finally, additional information on two important legal cases, previously mentioned.

Local actions: Establish Cooling Centers

As noted, high heat kills, more than any other extreme weather event. Although no one died in the Lake City during the 2021 heat wave, many were admitted to Kootenai Health hospital (Figures 11, 18 & 19) and visited its Emergency Department. Plus, there were significant increases in asthma, COPD and heat-related illnesses at Marimn Health, in Worley, just south (Figure 20). An estimated twenty people lost their lives in Spokane County. Across Washington, as many as 441 more people succumbed than expected, from June 26 to July 2 (Zhou, A., 2023).

Even without the historic heat wave in 2021, summers here are hotter than in the past. Four days in August 2023, for example, were intensely hot, topped by the record-breaking 103°F on the 16th. The summer of 2022 had five heatwaves, four in August.

While it's not news that our summers have gotten hotter, most people don't know what a cooling center is or how to find one. When I called Coeur d'Alene City Hall in early July 2022 to ask about them in town, they referred me, laughably, to the Idaho Dept. of Water Resources.

What is a cooling center? Any facility that provides air-conditioned shelter and water from the heat. In Kootenai County, churches and libraries have served as cooling centers. Only St. Vincent de Paul's homeless shelter in Coeur d'Alene has done so consistently, though, according to Kootenai County Office of Emergency Management (OEM) Preparedness Coordinator, Sarah Long.

According to Long, when dangerously hot weather approaches, OEM calls places that served as cooling centers the previous year to see if they would do so this year. Available facilities are posted on their website and on social media. For those with no internet, residents can sign up for emergency notifications by phone through Alert! Kootenai (www.kcsheriff.com/228/Alert-Kootenai).

What's most needed, though, is establishing a list of reliable facilities in each community. Calling senior centers, churches and recreation centers, in coordination with Long, is a project a small group of volunteers can do. That list should then be on the OEM website, with addresses and hours of operation, plus on a recorded OEM message. In addition, noted in local newspapers as well as known to hospitals, schools, social service agencies, libraries and realtors.

To maximize usage by low-income and high-risk individuals, centers ought to be places that are part of people's normal routine. That minimizes worries about strangers and being around screaming kids. Plus, have internet and activities for folks to do, plus some food, if possible. These recommendations come from a survey of Seattleites following the 2021 heat wave, noted in a *Grist* article (Bittle, 2023).

Protecting people from extreme heat is the highest priority, given it is most likely to harm, even potentially kill people. However, cooling centers should also provide relief from wildfire smoke, with the highest quality air filtration systems (HEPA filters). Currently, only schools and healthcare facilities use these more expensive air filtration systems, when conditions dictate.

With greenhouse gas emissions continuing to increase, at some point it's likely that people will need to be housed overnight, even multiple days, in especially dangerous conditions. We can't wait for people to die to develop this increasingly important resource. Cooling centers, thankfully, have been found to be life savers. Visting one on an extreme heat day reduces the risk of heat-attributable death by sixty-six percent (Bouchama, et al., 2007).

The elderly are the most vulnerable. Heat-related illnesses lead to thirteen thousand deaths each year and eighty percent of the people who die are over sixty. Nationwide, "nearly thirty percent of older adults live alone — in homes or apartments, by themselves, without spouses or in-home care. Isolation and lack of social connectedness translates into additional risk during disasters like heat waves" (Sightline, 2024). Nearly twenty-five percent of Kootenai County residents are over sixty; twenty percent in the Lake City (2020 U.S. Census Bureau).

The Building Threat of Wildfires

For many towns in the Inland Northwest, it's not a question of if, but when they're going to be hit by a big wildfire. Higher temperatures from climate change suck the moisture out of the ground, making it tinder dry and easily combustible. Fire suppression has provided a ladder for flames to reach high into the treetops as well, producing more intense, harder to manage blazes.

What's the level of risk we're facing for a major wildfire here in North Idaho? As noted previously (see Figure 16), a 2015 research paper concluded that we'll see a 300-percent increase in "megafires" over the next fifty years (NOAA, global science). A "megafire" is 12,000 acres or more.

High winds, during a prolonged drought in which large amounts of vegetation and forest soils become very dry can create a major wildfire. Such fires could destroy hundreds, if not thousands, of homes and businesses, as well as threaten many lives throughout the Inland Northwest.

Imagine if an iconic location such as Tubbs Hill in Coeur d'Alene were to go up in flames. It's not unheard of: fourteen smaller fires burned on Tubbs Hill during the summer of 2015, double the usual number (Oliveria, 2015). The lives and structures of neighbors and nearby businesses would be threatened. Smoke from the fire would exacerbate serious respiratory conditions in children and adults as well.

WHAT WE CAN DO TOGETHER

In addition, the loss of Tubbs Hill, an iconic landmark, would be a major cultural loss for the community. A hiking and swimming destination for generations and a major drawing card for tourists, its loss would exact an emotional as well as a financial toll on the town. Its restoration would take decades as well.

Protection From "The Big One"

What can our community do to protect itself from wildfires? We'll begin with a Forest Service project northeast of Hayden Lake that started in January 2022 ((U.S. Forest Service, 2022).

The 52,600-acre Honey Badger Project, named for the Honey and Badger mountains on its east side, extends from just south and east of Hayden Lake to the north, nearly to Farragut State Park. The largest project ever undertaken in the Idaho Panhandle National Forest, it involves fifty miles of road construction, 245 miles of new recreation trails, and more than 12,000 acres of logging.

Two major studies of the project area, in 2017 and 2018, found the forest with poor resistance to insects, diseases, drought and fire. Root disease was widespread, along with blister rust and a high danger of mountain beetle infestations among the less diverse types of trees. The project recommended an infusion of healthy western larch, western pine and ponderosa pine, which are much more resistant to disease and insect infestations.

Twelve thousand acres of forest were proposed for commercial timber harvest. In addition, planting of tree saplings in the open spaces provide a much healthier forest over time. A healthier forest is less likely to catch on fire than a dying and insect-infested one.

A sustainable trail system (especially in the Canfield Mountain area) is to be provided as well, along with reduced sediment discharge to streams and restoration of passage of aquatic organisms.

Now to Avista Utilities' efforts to minimize wildfire risk to its power distribution and transmission system. Following The Camp Fire that destroyed Paradise, California in 2018, Avista (and all utilities nationwide) developed an enhanced ten-year "Wildfire Resiliency Plan" in 2020.

Forty percent of Avista's service territory is in the interface between forest lands and human development, called the Wildland Urban Interface (WUI). Homes and businesses located near the WUI are most at risk from the impact of wildfires.

No part of the Lake City is in Avista's identified WUI area. However, extensively forested areas and towns north, south and east of town are. This puts them in the bulls-eye of high wildfire danger. See Avista's "Wildfire Resiliency Plan" on its website for those high-risk locations (Avista Utilities, 2020).

Over a ten-year period, Avista is investing $270 million to cut back vegetation near power lines and replace wood cross arms in WUI areas with fiberglass units. Also, to decrease forced power line outages, Avista is replacing wood poles with steel poles at "high value" locations (e.g., highway crossings, corner poles and heavy equipment poles).

In early May 2024, Avista upped the ante, announcing that rolling blackouts will now be part of their plan to combat area wildfires. The shutoffs would occur when fire conditions are the most dangerous. Avista would use forecasts to give alerts that begin with a "watch" as much as a week out, changing to a "warning" within a couple of days and an "imminent" announcement if utility managers plan to cut power (Clouse and Stephens, *The Spokesman-Review*, May 8, 2024).

During the announcement, Cheney Fire Chief Tom Jenkins "noted that Predictive Services at the National Interagency Fire

WHAT WE CAN DO TOGETHER

Center ranks northeast Washington in its top 10 most hazardous places to live because of the danger of large wildland fires" (Clouse and Stephens, *The Spokesman-Review*, May 8, 2024).

On a larger scale, we need to end future large-scale development in the WUI. The costs are too high not to. Our safety as existing property owners should not be second to the short-term profits of outside developers.

Montecito, California's Success Story

Montecito, California is an unincorporated, seaside community of 9,000 near Santa Barbara, CA. It's home to Oprah Winfrey and Ellen DeGeneres and other celebrities, as well as low and middle-income people, many of whom work in the service industry there.

A 2020 webinar sponsored by republicEn, "Living with Wildfire in the Era of Climate Change" (republic.En, 2020), featured Dr. Crystal Kolden, Ph.D. She described a case study of Montecito she conducted before, during and after the Thomas Fire of 2017. The study revealed how the city limited the fire's impact on Montecito and the surrounding area (Kolden, 2019).

Throughout its history, Montecito had been the victim of numerous fires. Over a twenty-year period, beginning in the mid-1980s, the Montecito Fire Protection District educated community members about the dangers of wildfire there and coordinated a city-wide effort to protect the city from the Big One (and other smaller conflagrations).

The result: the loss of only seven structures in the massive 2017 Thomas Fire, which burned nearly 282,000 acres and caused over two billion dollars in damages. Effective protection, then, can be done.

Three aspects of Montecito's fire protection plan grabbed my attention. First, the importance of educating, and directly involving, community members in developing and implementing protection measures prior to the fire. Second, empowering firefighters to be able to fight and escape from the fire. Third, the development and following of evacuation plans.

In 1990, the local Fire Protection District identified areas of high fire hazard. Wildland Fire Specialist positions were created to facilitate programs to reduce wildfire vulnerability, hazardous fuels, increase community engagement, code enforcement and conduct defensible space surveys.

Between 1999 and 2018, the Montecito Fire Protection District spent approximately $1.76 million ($2 million adjusted for inflation to 2018) on wildfire vulnerability reduction activities.

The Specialists reduced fuel adjacent to homes and roads and decreased the flammability of homes and infrastructure. They engaged residents individually, to develop a relationship and trust. These relationships were critical to overcome language barriers, distrust of government entities and to develop evacuation plans for those unable to drive themselves or those dependent upon others for transport.

The Specialists also found solutions that were inexpensive, free or found grant money to support individual properties or subdivisions. The Wildland Fire Specialists also facilitated helping residents evacuate in a timely manner. Residents had somewhere pre-determined to go, and with a 'go-bag' or checklist of critical medicines or valuables.

Some homes used fire-resistant materials to reduce the potential of the home becoming fuel for the fire. Others removed both live and dead trees, limbed trees or replaced them with less flammable landscaping (e.g., replacing native vegetation with succulents or herbaceous flora).

WHAT WE CAN DO TOGETHER

Property owners were also incentivized by the fire protection district to have contractors chip and remove woody materials. In a partnership with the public utility, Southern California Edison, many dead and dying trees near power infrastructure were removed.

High-resolution paper maps were printed and pre-packaged in a portable file box, allowing firefighters not familiar with the area to navigate safely and quickly and find pre-designated water sources. Several new fire codes were implemented as well. One required wider drive ways to improve evacuation and allow large firefighting equipment access to homes, turn around and get out quickly, if necessary.

Evacuation orders were issued five days before the fire reached the eastern edge of Montecito. Nearly all of the residents heeded the warning. This allowed fire suppression resources to prepare properties and move freely on the low-visibility roads without the hazard of civilian vehicles.

High winds, with sustained speeds exceeding twenty-nine miles per hour and gusts exceeding sixty-seven miles per hour, flung embers a quarter mile ahead of the main fire. Due to those winds, no helicopters or retardant tankers were able to be used.

Outreach and public education efforts ensured that sensitive populations had solutions to meet their needs. Fire personnel knew where individuals lived, who might require extra support during evacuation. Without that knowledge, many older individuals and children had died in earlier fires for lack of transportation to evacuate.

How might the strategies in Montecito work in Coeur d'Alene? Since community involvement is critical in protecting a community from wildfires, it's important to understand the type of community we live in.

In firefighting terms, Coeur d'Alene appears to be a High Amenity and High Resource (HAHR) Wildland Urban Interface Community. Leavenworth, Washington is as well.

HAHR communities have small or mid-sized populations, subdivisions or lot structures with homeowners' associations, and residents with higher expectations for community and firefighter services (e.g., road maintenance, parks, fire protection). Their economies are also more likely to be service or recreation based and centered on exceptional natural beauty.

These communities also tend to have "relatively high trust in government agencies managing nearby lands." However, their "... direct experience with fire in the landscape often tends to be low, given the relatively large proportion of former urbanites...." (Carroll & Paveglio, 2016).

Therefore, it's hoped public engagement with the U.S. Forest Service, the Bureau of Land Management and other governmental or non-profit agencies here would be less of a problem than in other, more suspicious or distrustful communities. Few language barriers would be helpful as well.

However, a great deal of education would likely be needed for those used to living in urban areas, to reduce woody materials and vegetation near their homes and along nearby roads.

Results from Montecito and other wildfire prevention success stories could be used to begin to coordinate a plan to protect Inland Northwest communities like the Lake City from wildfires.

Developing such a plan would likely include representatives from the local fire protection districts, the U.S. Forest Service, Bureau of Land Management, Kootenai County Emergency Management, the Coeur d'Alene Tribe, University of Idaho faculty, and Kootenai Environmental Alliance, among other groups. Applying for grants

WHAT WE CAN DO TOGETHER

and developing a relationship with members of the Montecito Fire Protection District (or other communities) would be valuable as well.

For example, Kootenai County can apply for funding with the Forest Service to protect communities from wildfires. A second round of grants, totaling $800 million, from the bipartisan 2021 Infrastructure law, are available. A Community Wildfire Protection Plan is developed first, then money provided to remove vegetation around homes and structures, establish evacuation corridors, create fire breaks in surrounding forests and educate residents about the risk (Brown, 2023).

We also need to learn to use fire as a tool to maintain the long-term health of our forests. An article by *The Nature Conservancy* noted that:

> Fire has been essential to the health of forest ecosystems for millennia. Untamed and frequent burns, sparked by lightning, shaped the diversity of life—nearly 80 percent of the native vegetation in North America evolved with fire. Since time immemorial, Indigenous peoples in North America skillfully used fire to manage the land....

Further, the article notes forests can be restored by: "... thinning smaller diameter trees and using controlled burns that reduce highly flammable fuels.... In other places, restoration may include reducing the density of trees and creating a network of openings in the forest canopy ... giving ponderosa pine and other fire-resistant trees more room to grow" (*The Nature Conservancy*, 2021).

Finally, important action by the Biden administration late in 2023. The Forest Service plans to prohibit cutting down old-growth trees, which store vast amounts of carbon, helping to limit climate

217

change. The trees also "provide an essential habitat for hundreds of species of wildlife and are more likely to survive wildfires." This would be the first time the Forest Service would "simultaneously revise all 128 of its forest plans, which dictate how all 193 million acres of forests and grasslands are managed" (*The Spokesman-Review*, December 20, 2023).

Communities Sign the Plant Based Treaty

What's a Plant Based Treaty? Launched in 2021, the non-binding treaty encourages people to eat more plants (not stop eating meat). Some twenty-five cities have signed on, including Boynton Beach, Florida, Edinburgh, Scotland and Los Angeles. In addition, the treaty "presses for no new land be cleared for animal agriculture and that ecosystems and forests be restored" (Buckley, *New York Times*, 2024). Volunteers could research the treaty, for possible future city council adoption.

State Actions

I attended the first-ever youth constitutional climate trial, previously mentioned, from June 12-20, 2023, in Helena, Montana. Sixteen Montana youth sued the state for denying their Montana-guaranteed constitutional right to a "clean and healthful" environment. The young plaintiffs testified to the personal harms they've suffered at the hands of Montana's fossil-fuel- based energy system.

The children described being afraid of the loss of "beautiful places" where they grew up, and the glaciers in Glacier National Park. Wildfire smoke threatens their still developing lungs, yet staying indoors felt like prison for the youth who love Montana's outdoors. The first fire of the current season, which doesn't usually begin until May, was on December 1st, it was noted.

WHAT WE CAN DO TOGETHER

Snow melt is two weeks earlier than normal, leaving diminished and hotter river flows in August. Fish are dying. Elk, bird and deer populations have diminished due to drought, the children and expert witnesses testified.

Why this loss of their home? Carbon dioxide, and other greenhouse gas emissions from coal-and natural gas-fired power plants and oil facilities in Montana that are "disproportionally large for its population," according to expert witness and Stanford University civil and environmental engineering professor Mark Jacobson. Montana has the sixth-highest CO_2 emissions in the country, higher than more than 100 countries, despite a population of only one million.

An expert on energy transitions, Professor Jacobson noted that the Big Sky state could meet one hundred percent of its energy needs by 2050 through water, wind and solar, plus energy efficiency improvements. He emphasized that a clean, renewable energy system would be "technically and economically feasible and economically beneficial."

In addition, Montana has the third largest wind energy resource in the country—330 times more than needed to meet its energy needs in 2050, he said.

Jacobson offered this as well: electrifying Montana's energy system would reduce its overall energy demand by sixty-one percent. How, I wondered? By eliminating the need and expense of mining for fuels, either coal, oil or natural gas. Instead, improved energy efficiency, existing hydropower, plus wind turbines and solar panels would provide all the electricity the state needs in 2050 (Expert witness testimony of Mark Jacobson, *Held v State of Montana*, June 16, 2023). Montanans' home needn't be sacrificed.

The current system creates mounting health, recreational and climate costs. Money spent on aging and inefficient coal-fired power plants would no longer be incurred, as well as to extract, transport and burn coal, oil and natural gas. Home could not only be saved but restored over the long-term.

Here in Idaho, we're in a better situation than Montana. Not only does the Gem State have few fossil fuel reserves, but in 2022 renewable energy generated seventy-five percent of our in-state electricity. Most comes from hydropower, although we burn natural gas, as well as use wind, solar, biomass and geothermal resources (U.S. Energy Information Administration, 2019-2022).

If Idaho's energy needs in 2050 were powered by wind, water and solar energy, The Solutions Project, developed by Professor Jacobson, projects the following Energy Mix:

- 35% Onshore Wind
- 17.8% Solar Plants
- 15% Geothermal
- 15% Hydroelectric
- 10% Concentrated solar plants
- 4% Residential rooftop solar
- 3.2% Commercial & government rooftop solar

Meeting our state's energy demands by these safe, renewable energy sources, it's estimated, would reduce the overall energy demand by thirty-seven percent. There would be a $2.42 billion health cost savings per year and 219 lives previously lost to air pollution would be saved each year. In addition, the transition would pay for itself in as little as 4.3 years from air pollution and climate cost savings alone.

Each person would receive $188 energy cost savings as well. Adding health and climate cost savings would result in $5,468 savings per person (The Solutions Project).

The Power of Movements

> "Nonviolence is a powerful and just weapon which cuts without wounding and ennobles the man who wields it. It is a sword that heals."
>
> Dr. Martin Luther King

When asked what people can do to make a difference with climate change, environmentalist, advocate and author Bill McKibben said "Become more than just an individual." A powerful way to do this is to join a movement with others.

At a weekend training in Seattle in 2015, I learned how to engage in non-violent, direct action to oppose the Keystone XL pipeline (carrying heavy oil from northern Alberta to Gulf of Mexico terminals). Lisa, my wife, expressed concern for my safety and employment if I were to join an action. However, I felt strongly that the issue was important enough to risk both.

It was comforting to read that the American Counseling Association valued participation in non-violent activism. The following year I signed an online pledge of direct action, if necessary, at the Keystone XL pipeline site in North Dakota, risking arrest. As it turns out, I wasn't called. President Biden withdrew the pipeline's permit in 2021 and the project was terminated by its developer.

Before the pandemic, I participated in a few local actions with 350.org, an international climate advocacy group, founded by McKibben.

From an early age, I've been inspired by the courage of Martin Luther King. He and his followers faced water cannons, attack dogs, club-wielding police and vicious mobs in their fight for civil rights. I also admired Mahatma Gandhi's efforts to obtain India's independence from Great Britain, and abolitionists such as Frederick Douglass and Harriet Tubman.

I was surprised that the effectiveness of mass movements has been studied. Movements, research indicated, especially those involving nonviolent protests, often produce serious political change. The phenomenon is called the 3.5 percent rule: when 3.5% of a population actively engages in a movement, success has been nearly guaranteed.

Nonviolent protests are twice as likely to succeed as armed conflicts, researchers Erica Chenoweth, a political scientist at Harvard University, and Maria Stephan, a researcher at the International Center of Nonviolent Conflict found. Data from 323 violent and non-violent campaigns, from 1900 to 2006, were analyzed and published in 2011's *Why Civil Resistance Works: The Strategic Logic of Nonviolent Conflict.*

I wanted to know what about non-violence works. First, strength in numbers, the researchers said. Nonviolence attracts more and a wider range of people than violent protests. Second, nonviolence maintains the high moral ground. Third, it's easier to discuss the resistance openly when it's nonviolent (Chenoweth & Stephan, 2011). Finally, how many people would be 3.5 percent of the U.S. population: currently, about eleven million people.

More recent research, however, finds the critical tipping point for changing everyone's behavior is a committed minority of twenty-five percent (Centola, 2021). Amassing large enough numbers to affect significant change takes years.

There are many national and international actions to oppose new fossil fuel projects. If any actions were nearby, or were morally compelling, I would participate in-person. It's vital, as you'll read in the next section, to keep some eighty percent of the world's remaining fossil fuel reserves in the ground.

Keeping 80% of Remaining Fossil Fuels in the Ground

Actions toward a net-zero emissions future are no longer being driven by environmentalists alone. Huge asset management, banking and insurance companies have large sums of money in the game. Why? An increasing number of investors in Big Oil are worried a large portion of their fossil fuel assets may become stranded, unable to be burned, due to governments legislating to curtail climate risk.

The push for keeping as much as eighty percent of the remaining fossil fuels in the ground is gaining ground. The big players show up when huge sums of money are at stake. Hopefully, they can turn the tide on global warming.

According to the founder of the Carbon Tracker Initiative, Mark Campanale, as much as twenty-five to thirty trillion dollars of fossil fuels will need to be left in the ground (i.e., as stranded assets) to meet climate change requirements.[3] Another estimate: the lost financial value to the fossil fuel industry and certain countries (Saudi Arabia and Venezuela, most notably, which are highly reliant on oil and natural gas revenues for their economies) could hit $100 trillion (*The Guardian*, 2018).

I first heard of the need to leave eighty percent of the world's remaining fossil fuels in the ground in a summer 2012 *Rolling Stone* article by Bill McKibben. As the headline screamed, it was scary news: "Global Warming's Terrifying New Math" (McKibben, 2012).

McKibben laid out three numbers in the article:

---

[3] According to its website, the Carbon Tracker Initiative is "an independent financial think tank that carries out in-depth analysis on the impact of the energy transition on capital markets and the potential investment in high-cost, carbon-intensive fossil fuels."

- **2° Celsius (3.6°F.)**—the maximum safe global temperature, noted by the 2009 Copenhagen climate accord (the 2015 Paris Climate Agreement revised the safe limit to 1.5°-2°C. (2.7°-3.6°F.).
- **565** Gigatons — how many more gigatons (a billion tons) of CO_2 can be poured into the atmosphere by mid-century to have some reasonable hope of staying below 2°C.
- **2,795** Gigatons — the amount of carbon contained in the proven coal, oil and gas reserves of the fossil-fuel companies, and oil-dependent countries like Venezuela and Kuwait).

Notice, McKibben wrote —**2,795 is five times larger than 565**. In other words: "We have five times as much oil and coal and gas on the books as climate scientists think is safe to burn. We'd have to keep 80% of those reserves locked away underground to avoid that fate." "If you told Exxon or Lukoil that, in order to avoid wrecking the climate, they couldn't pump out their reserves, the value of their companies would plummet."

> "...if you paid attention to the scientists and kept 80 percent of it underground, you'd be writing off $20 trillion in assets. That's a catastrophic carbon bubble. *Either you can have a relatively healthy planet or a healthy balance sheet for fossil-fuel companies. But, you can't have both*" (McKibben, B., 2012). [author's emphasis]

Over the past ten years, there's been some refinement in that 80% figure. A September 2021 study calculates that: "... nearly 60% of the world's oil and gas reserves and 90% of the coal reserves need to stay in the ground by 2050 to meet climate goals of the Paris

Climate Agreement. Those limits would give the world a 50-50 chance of limiting global warming to 1.5° C (2.7°F) compared to pre-industrial times...." (Costley, 2021).

The University College London researchers also noted that global production of fossil fuels needs to begin to decline immediately, at an average annual rate of around three percent through 2050 (Costley, 2021). As noted earlier, the U.S. is to do one-quarter of that, the European Union one-half.

NATIONAL ACTIONS

Pricing Carbon

In 2019, the non-partisan Citizens' Climate Lobby (CCL) introduced the Energy Innovation and Carbon Dividend Act in the U.S. House of Representatives. The Act calls for an increasing fee over time to be placed on fossil fuels at the source (i.e., the mine, well head or port of entry). One hundred percent of the revenue would be returned to U.S. households as a dividend.

The price would start at fifteen dollars per ton of CO_2 embedded in the fuel (coal, oil or gas) and go up ten dollars per ton every year. A border adjustment is a tax on imports and rebates on exports which accounts for differences in carbon pricing across differing countries. Such a tax would discourage businesses from relocating to where they can emit more CO_2 without paying the tax.

How effective would it be? According to a study by Regional Economic Models, Inc., greenhouse gas emissions would be cut by fifty percent in twenty years. In addition, it would create 2.8 million jobs, boost the country's GDP and save 200,000 lives (Citizen Climate Lobby, of which this author is a member).

The legislation had ninety-five co-sponsors, more than any other carbon pricing legislation in U.S. history. However, in a Republican-majority House of Representatives, it almost certainly will not be voted on. Worldwide, thirty-nine countries, including Canada, already use carbon pricing (The World Bank, Carbon Pricing Dashboard).

Citizens' Climate Lobby is also pushing to speed up assessing the environmental impact of clean energy projects, and where approved, build and connect them to the electrical grid. Without that reform, it says eighty percent of the potential carbon pollution reduction from the Inflation Reduction Act will be lost. Now it takes an average of 4.5 years for federal agencies to complete the environmental impact assessments for major energy projects (CCL email, Sept. 5, 2023).

In late April 2024, Department of Energy rules ought to help with this backlog and bottleneck of grid transmission projects. DOE will now be the single coordination between agencies and a cap for the federal permitting process was set for two years. This should cut the average wait time for permit approval in half (Allsup & Jenkins, 2024).

Streamlining this process is crucial. Some 2,600 gigawatts of planned power projects were seeking to connect to the U.S. grid at the end of 2023. This represents twice as much as the country's existing capacity, according to a report by Lawrence Berkeley National Laboratory. Solar, wind and storage projects make up ninety-five percent of the capacity in the interconnection queue (Rand et al., 2024).

To return to Citizens Climate Lobby, Coeur d'Alene has a chapter, led by Dave Muise. Sandpoint and Moscow have chapters as well. To join, and support the national legislation, go to www.climateeducation.org.

Juliana v. U.S. climate lawsuit

> "Exercising my reasoned judgment, I have no doubt that the right to a climate system capable of sustaining human life is fundamental to a free and ordered society."
>
> U.S. District Judge Ann Aiken
> Juliana v. U.S. court case,
> 2016 decision

> "They know that once you enter that courtroom and your witnesses take the oath to tell the truth and nothing but the truth and the facts are the facts and alternative facts are perjury. *And so all of these claims and tweets about climate change not being real, that doesn't hold up in a court of law.*"
>
> Julia Olson,
> executive director,
> Our Children's Trust
> [this author's emphasis]

The Juliana v. U.S. court case is not an ordinary lawsuit. Indeed, it's a landmark climate change case. Twenty-one youth are suing the federal government for depriving them of their constitutional rights to life, liberty and property (due to the impacts of climate change where they live) and failing to protect essential public trust resources (e.g., the atmosphere).

For the first time in a major court of law, the U.S. government has admitted climate change is driven by the burning of fossil fuels.

On March 3, 2019, Julia Olson, Executive Director of Our Children's Trust, the non-profit public interest law firm representing the young plaintiffs, was interviewed on *60 Minutes*. Olson said that the Trump administration admitted: "… that the government has known for over fifty years that burning fossil fuels would cause climate change. And they don't dispute that we're in a danger zone on climate change. And they don't dispute that climate change is a national security threat and a threat to our economy and a threat to people's lives and safety. They do not dispute any of the facts of the case (CBS News, March 3, 2019)."

Olson continued:

> The government has been forced to admit that:
>
> … global carbon dioxide concentrations reached levels unprecedented for at least 2.6 million years…. That climate change is increasing the risk of loss of life and the extinction of many species and is associated with an increase in hurricane intensity, the frequency of intense storms, heavy precipitation, the loss of sea ice and rising sea levels. And the government acknowledges that climate change's effects on agriculture will have consequences for food security.

Olson then added: "It's really the most compelling evidence I've ever had in any cases I've litigated in over 20 years" (CBS News. 2019, March 3).

I continue with excerpts from the 60 Minutes interview, conducted by Steve Kroft: Olson: "The lawsuit … argues that the government has failed in its obligation to protect the nation's air, water, forests and coastlines."

Kroft: "Why is the federal government responsible for global warming? I mean it doesn't produce any CO_2. How are they causing it?

Olson: "They're causing it through their actions of subsidizing the fossil fuel energy system, permitting every aspect of our fossil fuel energy system, and by allowing for extraction of fossil fuels from our federal public lands. We are the largest oil and gas producer in the world now because of decisions our federal government has made."

Acknowledging that the U.S. is not solely responsible for global climate disruption, Olson added that the U.S. has put twenty-five percent of the total greenhouse gas emissions into the world's atmosphere over the decades (CBS News, 2019, March 3).

Ms. Olson stated that she knows the government will lose if it goes to trial. The Trump administration tried and failed twice before the U.S. Supreme Court and three times before the Ninth Circuit Court of Appeals to get the case dismissed. The Obama administration opposed the case in its early stages as well. Initially, the oil and gas industry was a defendant, but afraid its records could be used against it, withdrew from the case.

The Biden administration, unfortunately, has maintained the same legal defense as the Trump administration. In March 2021, Our Children's Trust filed a motion with the U.S. District Court in Eugene, Oregon to not require the government to develop a climate recovery plan, an aspect of the case objected to by the 9th Circuit Court.

That motion was heard in June, 2021 by Judge Ann Aiken (whose landmark judicial ruling is highlighted in this chapter's opening). Two years later, Judge Aiken ruled in the children's favor. After the government attempted to have the case dismissed, on December 29, 2023, she ruled that the case can go to trial. The judge's words, in her ruling, highlight what's at stake in this groundbreaking case:

"This catastrophe is the great emergency of our time and compels urgent action. As this lawsuit demonstrates, young people — too young to vote and effect change through the political process — are exercising the institutional procedure available to plead with their government to change course."

In February 2024, the Department of Justice filed a rarely used legal maneuver to keep the case from going to trial. This is its twenty-second attempt to squash the case, the most ever in any federal judicial case (People vs Fossil Fuels email, February 26, 2024). On April 19, Judge Aiken told the U.S. Court of Appeals that the government's arguments were not valid and the case should go to trial (OCT email, April 22, 2024).

Unfortunately, on May 1, the Ninth Circuit Court of Appeals ruled against the children again, dismissing the Juliana v. United States case. The court said that the declaratory relief Our Children's Trust sought — stating what the law is, rather than what the government needs to do to remedy the issue — was "not substantially likely to mitigate the plaintiffs' asserted concrete injuries" (Showalter and Rasche, *The National Law Review*).

The case is not closed, though. A petition for the full nine-judge panel to hear the case is due by June 17th. In addition, Our Children's Trust is seeking amicus briefs by the nation's legal experts, scholars and members of Congress to "address the severity of the decision on justice, our nation's children, and the climate emergency" by June 27th (OCT email, June 2, 2024).

Earlier, in 2023, there were important court wins. In November, three Canadian Federal Court of Appeals justices unanimously allowed a similar case to go to trial, after the complaint is modified (Our Children's Trust email, December 14, 2023). A Virginia case is currently before the court of appeals (OCT, 2023) and the Navahine

F. v. Department of Transportation youth state constitutional climate trial in Hawai'I is scheduled for June 24, 2024.

In December, eighteen California youth filed a lawsuit against the U.S. Environmental Protection Agency for violating their constitutional rights to a healthy life (OCT, 2023). Our Children's Trust is arguing climate cases in Mexico, Portugal and other countries as well.

Pressure on state and national governments is ramping up. If the reader wishes to support such efforts, they can do so at donate.ourchildrenstrust.org.

Increasing Renewable Energy

In 2022, renewable energy was the fastest growing power source in the U.S. Solar power represented the fastest-growing electricity source (Sylvia, 2022). In 2023, renewable energy generated twenty-one percent of the country's electricity. Natural gas generated forty-three percent, coal sixteen percent and nuclear nineteen percent (Energy Information Agency, 2024). The price of solar panels dropped eighty-five percent between 2010 and 2020 (Yale Climate Connection, 2023).

Locally, the Coeur d'Alene Tribe has invested in renewable energy. Its first solar photovoltaic system was installed in 2016, then another in 2020. In 2023, a thirty-six kilowatt system was installed at the Marimm Coeur Center (*Climate Change on the Coeur d'Alene Landscape*, 37).

On Earth Day 2024, the EPA selected sixty programs nationwide to receive funding from the $7 billion Solar for All Program for low-income families and disadvantaged communities. In Idaho, $56 million will be administered by the Bonneville Environmental Foundation for single and multifamily residential solar projects. Twenty-five percent of the funds will be allocated to Idaho tribes. The

Coeur d'Alene Tribe expects to receive awards in September, with bids and installations to occur late in 2025 (Laumatia email, 2024).

Renewable energy is the future. In 2020, when oil prices collapsed during the pandemic, approximately 160,000 oil and gas company employees were laid off. Many have been hired by renewable energy businesses. Employment in the oil and gas industry declined "over 20%.... By comparison, employment in wind energy grew nearly 20% from 2016 to 2021, to more than 113,000 workers."

Why this big switch to renewable energy as an employer?

> In more than a dozen interviews, energy workers and executives said they had switched…because they felt the oil and gas industry's best days were behind it. Others said they were no longer willing to tolerate the extreme ups and downs of oil and gas prices, and the accompanying cycles of rapid hiring followed by crushing layoffs. Many said concerns about climate change…were a factor in their decision (Krauss, 2023).

In addition, renewable energy is safe, clean and eliminates the constant need and expense of mining for fuels, noted earlier (pages 217-218). Indeed, there are no fuels with renewable energy; it never runs out. Plus, overall energy demand is reduced, as are health costs (fewer air pollution cases) and climate costs (e.g., reduced heat- and smoke-related respiratory conditions).

Regional contributions

Eastern Washington is contributing to important solutions as well. A Silicon Valley startup, Twelve, broke ground on a plant in Moses Lake, Washington in July 2023. It will produce sustainable aviation fuel. The fuel is key to decarbonizing the aviation industry (Gates, D., July 2023). The plant could be in production by spring 2024 (Schweitzer, *Basin Business Journal*, 2023) and eventually employ 500 workers (KREM-2 News, 2023).

Named for the most common isotope of carbon, Twelve, will use electricity from solar, wind or hydropower to attract hydrogen from water. It is then combined with CO_2 captured from the air to produce hydrocarbons like those derived from oil. The aviation fuel is commonly called e-fuel. Twelve calls its product E-Jet. It's expected to have ninety percent lower lifecycle emissions than regular jet fuel (Gates, D., June 2023).

Moses Lake will also be at the forefront of building better batteries for electric cars. It will utilize $100 million from federal funds, and promises 500 new jobs. Sila Nanotechnologies will harness the power of silicon anodes for commercial use in lithium-ion batteries. Anodes are the negatively charged components of batteries used in products such as cellphones and electric vehicles.

Using silicon can result in significantly higher energy storage, longer battery life and the ability to generate ten times more charge than graphite. In addition, graphite is a national security risk, since about ninety percent of it is currently mined and processed in China (Stephens, *The Spokesman-Review*, 2023).

Solar Dominance Hypothesis

I hadn't heard of this solution until I came across a YouTube video about it in 2020.[4] One of its major proponents, Tony Seba (Stanford professor and Silicon Valley entrepreneur) claims the world could be 100% powered by non-fossil fuel sources *by 2030*. Those sources

---

[4] "Unstoppable — The Solar Dominance Hypothesis," Feb. 16, 2020.

would include solar, wind, geothermal, tidal, a little nuclear and energy efficiency measures.

Seven years seemed to be far too optimistic, though, both for Amory Lovins[5], who explored the topic in a video and the following discussion, and to me. A more plausible scenario would be to have half the world's power generated by renewable sources by 2030.

The concept hinges on how quickly people adopt new technology and the dropping prices of those renewable sources. Consider the smartphone, Lovins said. It was introduced to the world in 2007. By 2017, just ten years later, there were 2.5 billion users, one-third of the world's population. This is a much quicker adoption rate than for the telephone, which took decades to reach saturation in the market. I saw this first-hand while visiting Peru in 2019. Despite widespread poverty, cell phones were used by everyone.

Ms. Lovins noted there are four drivers of the Solar Dominance Hypothesis:
1. Falling costs of solar electricity. Since 2000, solar capacity has doubled every two to three years; each doubling reduces the installed price of solar per megawatt (one million watts) by twenty percent.
2. Falling costs of solar storage. Lithium-ion battery prices are falling at about the same rateas solar.
3. Adoption of the electric car. Sales surged, then slowed recently.
4. Adoption of the driverless car.

What's happened here in the U.S. since that 2020 video? It took four years — from 2018 to 2022 — to double installed photovoltaic capacity. Lithium-ion battery prices fell fifty percent over four years, until 2019, and continue to fall more slowly. As previously described,

---

[5] American environmentalist, author, sustainable development proponent and co-founder of Rocky Mountain Institute.

adoption of the electric car has surged, but slowed recently. Adoption of the driverless car is progressing.

Update the Transmission Grid

The country's electric grid is more than 100 years old. It's in need of both repair and upgrading. The upgrading would move renewable energy resources to where they're needed and greatly improve the grid's efficiency and responsiveness. Such steps are essential to meet President Biden's goal of having a carbon-free electric generating capacity in the U.S. by 2035.

Grid updates make use of Grid Enhancing Technologies (GET)—hardware or software that increase capacity, efficiency and/or reliability of transmission facilities. A 2021 study found that GET adoption in Kansas and Oklahoma can deliver twice the renewable energy capacity. It also saves $175 million in production costs, while mitigating three million tons of annual CO_2 emissions. Both short- and long-term jobs would be generated as well (Unlocking the Queue, 2021, The Brattle Group).

Potential nationwide benefits include twenty million tons of carbon emission cuts equal to taking twenty million cars off the road. Production cost savings could be more than five billion dollars. Tens of thousands of local construction jobs and thousands of long-term, high-paying jobs could be created (Unlocking the Queue, 2021).

In April of 2021, the Biden administration released more than eight billion dollars to begin the grid update. Developers will use the money to improve the grid's ability to carry renewable energy from the windy, sunny plains to large metropolitan areas. To assist this process, the administration is pledging to help speed the siting and permitting of transmission projects by facilitating the use of public highways and other transportation rights-of-way.

The nonprofit organization, Americans for a Clean Energy Grid, dedicated to modernizing and expanding North America's high voltage grid, released a report in April 2021. It listed twenty-two shovel-ready transmission projects that could boost the use of wind and solar energy and generate 1.2 million jobs. The cost to build those transmission lines: thirty-three billion dollars.

If all twenty-two projects are completed, the 8,000 miles of wires could enable a fifty percent increase in the amount of wind and solar on the U.S. grid. However, those projects represent just ten percent of the transmission investment needed to completely decarbonize the power sector, the report said.

The report's co-author and founder of Grid United, Michael Skelly, noted: "They [the projects] can be underway in the next 12 to 36 months, if we can just get a little bit of a push to get those projects over the top" (Goggin, et al., 2021).

A much larger push was provided through passage of the federal Bipartisan Infrastructure Bill. The bill authorizes $65 billion to rebuild the electric grid, a good down payment towards decarbonizing the power sector (Evergreen Action & A Matter of Degrees podcast, 2022). The Inflation Reduction Act provides for planning, negotiation, and construction of new transmission lines and corridors.

Expand Grid Storage

A May 2022 TEDxReno talk by Dr. Denis Phares, CEO of Dragonfly Energy, was informative. Dr. Phares emphasized the need to generate sufficient quantities of electricity through solar and wind energy to deal with the climate crisis. The problem, however, is the intermittency of both: the sun doesn't always shine and the wind doesn't always blow.

What's needed, and being developed by his company and others, is a grid storage battery. Not one based on the lithium-ion battery, which has too high of a flammability risk for grid storage. The lithium-iron-phosphate battery meets the three requirements for grid storage: 1) energy dense enough to take up little space, 2) cost effective over its lifetime—able to handle thousands of charge/discharge cycles, and 3) non-flammable.

Rather than using a liquid electrolyte, which reacts violently with oxygen in lithium-ion batteries, the lithium-iron-phosphate battery will use a solid-state electrolyte, he said. That same concept is being developed for electric vehicles. Phares did not indicate when his company's grid storage battery will come on the market (Phares, 2022).

Net-Zero Emissions in the U.S. by 2050

A number of studies indicate how (no longer if) our country can get to net-zero emissions by 2050. But first, what does net-zero emissions mean? It means releasing no more greenhouse gas emissions into the atmosphere than are permanently removed each year.

I'll reference two studies on how to get to net-zero emissions in the U.S. by 2050. The first: "Carbon-Neutral Pathways for the United States," was released in Nov. 2020. It noted that:

> ... increasing energy efficiency, switching to electric technologies, utilizing clean electricity (especially wind and solar power), and deploying a small amount of carbon capture technology, the United States can reach zero emissions *without requiring changes to behavior.* Cost is about

> $1 per person per day, not counting climate benefits; this is significantly less than estimates from a few years ago because of recent technology progress.... The lowest-cost electricity systems get >80% of energy from wind and solar power, but need other resources to provide reliable service.... Key tasks for the 2020s: increasing the capacity of wind and solar power by 3.5 times, retiring coal plants, and increasing electric vehicle and electric heat pump sales to >50% of market share (Williams, J.H., et al., 2020).

Thus, getting to net-zero emissions by 2050 can be done, primarily by clean, renewable power, through one of the eight pathways noted in this study.

It can also be done at no greater cost than we're spending on energy now, and with technologies we currently understand, says another study: Princeton's Net-Zero America Report. However, it emphasizes how the things we do must change dramatically.

Access to at least $2.5 trillion of upfront capital by 2030 is needed to greatly expand the building of solar panels and wind turbines. Also, 250 million or more plug-in or fully electric vehicles need to be on the road by 2050. Today there are about two million. These actions and what are called the six pillars of getting to net-zero emissions by 2050 are described in a YouTube video by Dr. Eric Larson, a co-leader of the two-year Princeton project (Net-Zero America Report, 2020).

In one of the five pathways studied, the High Electrification Scenario, 100% of the cars would be electric by 2050 and coal would no longer be used by 2030 for power generation. Also, to achieve net-zero emissions, about 500,000 to one million new energy jobs will be needed across the country in the 2020s alone, more than offsetting fossil fuel jobs lost (Seltzer, 2020).

WHAT WE CAN DO TOGETHER

Much of that new clean energy infrastructure will go into wind power in the Midwest and Texas, plus offshore wind off the East Coast. In addition, it was recently announced that more than five million acres of federal public land in five Western states, including Idaho, could be opened up to solar energy production. Combined with seventeen million acres in six other Western states, the plan is to meet President Biden's target of a one hundred percent clean electricity grid by 2035 (Baumhardt, 2024).

Permitting and siting clean energy projects face many challenges. Obscured views are a big concern in some areas of the country. In order to avoid lengthy state by state, parcel by parcel approvals of new transmission lines, it's best to build energy sources close to population centers and within ten miles of existing or planned transmission lines.

The Interior Department has a goal of approving permits for twenty-five gigawatts of renewable energy on federal land by 2025. However, some of the Princeton models propose nearly five times that amount on public land will be needed in the coming decades. Destruction of habitat of various species, such as the desert tortoise and sage grouse, by adding vast solar panel arrays or hundreds of wind turbines is an important consideration (Penney, 2021).

Carbon Capture and Storage

As a graduate student in the mid-1970s, I wrote a master's thesis on the permanent disposal of highly radioactive waste from U.S nuclear power plants in deep salt beds. The right site would be dry, geologically stable and seal like plastic around the radioactively hot canisters. However, a proposed repository in Nevada was nixed in the Obama administration, and nothing further has been done towards permanent nuclear waste disposal.

239

What's the connection to climate change? The fossil fuel industry, which has belched so much carbon into the atmosphere, is helping fund a frantic race to store liquid CO_2 in wells injected into the Gulf Coast. Basically, oil companies would collect their own greenhouse gases and pay a company to compress it into liquid and inject it into deposits of sandstone five to twenty-three million years old.

While this would allow oil companies to keep polluting, with minimal cost, the Intergovernmental Panel on Climate Change has affirmed that extensive long-term storage is likely necessary to meet any of its targets to seriously mitigate global overheating. Using carbon capture and storage, it's hoped, would buy time to decarbonize the economy. Eventually it could be fully powered by renewable sources of energy.

Some of the biggest names in the petroleum industry: Chevron, ExxonMobil, ConocoPhillips and BP, are jumping in. In August 2021, the first rights to carbon injection sites were awarded. A sixty-three-square mile lease, in a roughly 300-mile arc of the Gulf Coast, from Corpus Christi, Texas to Lake Charles, Louisiana were designated as carbon injection sites. Chevron announced it would invest fifty million dollars for half of the lease. The EPA, however, hasn't issued its first permit for large-scale commercial injection. Permit reviews are widely expected to take years, with an uncertain outcome.

What are the risks? As one geologist put it in a 2020 *Wired* article, people will "have to decide that the risks of CO_2 going into the atmosphere are more fundamental than the risks of CO_2 going into the ground."

While it's ironic that the industry that's served up fossil fuels for a century may play an "influential part in curbing greenhouse gas emissions," the article's author noted, it's also "a measure of how economic signals are changing in a part of the world that has long adapted the way it exploits its natural resources to meet shifting market demand" (Ball, 2022).

It will take an all-in approach by oil companies to make a dent in digging ourselves out of the hole their hydrocarbons put us in.

What About Nuclear Power?

Producing no greenhouse gas emissions, fifty-five nuclear power plants currently operate in the U.S. (none in Idaho). The Union of Concerned Scientists' report found more than one-third of those operating at the end of 2017—twenty-two percent of the U.S. nuclear power capacity—were either unprofitable or slated to close within the next ten years. However, replacing those plants with natural gas-fueled plants could increase the electric power sector's carbon emissions by six percent by 2035 (*Post* Opinions, 2019).

There are numerous ways to preserve existing nuclear power and boost investments in renewables and energy efficiency. States that provide subsidies to prevent uneconomical plants from closing abruptly should limit those subsidies and adjust them over time to protect customers. Also, provide them only to nuclear plants that meet or exceed federal safety standards, according to a *Washington Post* article.

In addition, plant owners ought to be required to develop worker and community transition plans to prepare for facilities' eventual retirement and decommissioning. Most reactors are scheduled to expire between 2030 and 2050 (*Post* Opinions Staff, 2019).

The closing of the huge Colstrip coal-powered plant, in Montana, has such a requirement.

Displaced workers had until April 30, 2021 to apply for grants through Puget Sound Energy.

In Nov. 2021 Bill Gates' company, TerraPower, announced Kemmerer, Wyoming will be the site for testing its prototype next-generation sodium fast reactor, with molten salt energy storage.

Kemmerer was chosen as the site of one of Wyoming's four retired coal-fired power plants (TerraPower, 2021).

It's important that next-generation nuclear is being developed because building new nuclear power plants is hugely expensive, a new report states: "The central issue with nuclear power is cost. Constructing nuclear power costs approximately $10,000 per kilowatt of capacity. This is ten times as expensive as solar and wind, which cost around $1,000 per kilowatt" (Dessler, 2024).

Now for two pieces of federal legislation, with increasingly important impacts on reducing climate change.

Infrastructure Investment & Jobs Act

The Bipartisan Infrastructure Law includes $108 billion for public transport, plus sixty-five billion dollars to upgrade the power infrastructure. This includes building thousands of miles of new, resilient transmission lines to facilitate expansion of renewables and clean energy. In addition, $7.5 billion to build a national network of 500,000 EV chargers by 2030 (White House, 2022).

Inflation Reduction Act

Some $350 billion of the Inflation Reduction Act (IRA) is an investment in climate, energy, and environmental justice. Estimated to reduce carbon dioxide emissions by forty percent by 2030, over 2005 levels, it has the potential to be transformational. Here are some of its major elements:

- $7,500 tax credit towards the purchase of designated electric vehicles for families making less than $150,000 per year; $4,000 for a used EV, all through the year 2032.
- Tax credits for low and middle-income Americans to purchase clean energy technologies, such as an EV, a heat pump or an induction stove (more efficient, with no residual heat since the air between the cooktop and pan never gets hot). Rebates will cover fifty to one hundred percent of the cost of installing such appliances.
- The IRA covers thirty percent of the installation costs, plus battery storage systems, and home energy efficiency improvements, as well as upgrades to heating and cooling systems.
- "In Idaho, more than 13,000 people are currently employed in the solar, wind, storage, and other clean energy industries. Over the next eight years, the IRA will create many new local jobs with an infusion of about $320 million for energy storage projects and large-scale clean power generation."
- Funding for investments in wildfire prevention efforts and tree planting projects that protect cities from extreme heat.
- Offshore wind tax breaks are extended for ten years. Ten billion dollars are invested in domestic wind technology production.
- Twenty-seven billion dollar climate bank created, to provide low-interest loans (Stokes, podcast, 2022).

To secure Sen. Manchin's (D-WV) yes vote on the IRA, however, more oil and gas leases, particularly in the Gulf Coast and Alaska, were allowed. While counterproductive to fighting climate change, the IRA is, all in all, transformational legislation. Major climate legislation finally crossed the finish line.

How effective has the Inflation Reduction Act been so far in reducing greenhouse gas emissions? Eighteen months after its passage, progress so far is "a little short" of the forty percent reduction by 2030 goal. What's most needed, the Rhodium Group's Trevor Houser says, is a "dramatic jump" in utility-scale geothermal, solar, wind and battery storage projects.

Expediting the permitting and siting of projects, and mobilizing the supply chain, will be crucial to building many more clean energy facilities (Brocious, 2024). Many more clean energy workers are needed as well.

Reducing Methane Emissions

In early December, 2023, the Environmental Protection Agency finalized a rule that will reduce methane emissions by nearly eighty percent over emissions without the rule. Specifically, the rule will phase out flaring on new natural gas wells and require methane leak-detection surveys and repairs. In addition, there will be a first-ever $1,500-per-ton methane emissions fee. Over the next fourteen years, the rule is expected to prevent the release of fifty-eight million tons of methane emissions (EPA rule, December 2, 2023).

Agriculture contributes significant methane emissions from cows' flatulence and belching. After other low-level, failed attempts, the Future of Food program at the Bezos Earth Fund states that now "more than US$100 million is flowing to develop methane-reducing technologies in livestock and rice" (Jarvis, 2024).

WHAT WE CAN DO TOGETHER

Kigali Amendment

In September 2022 the Senate ratified, with bipartisan support, this amendment to the Montreal Protocol. It will phase down the worldwide use of hydrofluorocarbons (HFCs), powerful greenhouse gases.

HFCs were developed to replace chlorinated fluorocarbons (CFCs), used as refrigerants and propellants in items such as air conditioners, deodorants and hair spray. CFCs were found to be seriously damaging the protective ozone layer from ultraviolet radiation (with the threat of enormous numbers of skin cancers, serious sunburns and significant crop reductions).

Now, with passage of this legally binding amendment, the consumption of their replacement, HFCs, are to be reduced by more than eighty percent by 2047. The impact on climate change? Estimates are that up to a 0.5°C (0.9°F) increase in global temperature by the end of the century could be avoided. Also, eliminating CFCs is expected to restore the stratospheric ozone layer by 2065 (U.S. Secretary of State, 2022).

INTERNATIONAL ACTION

Urgent Methane Emissions Reductions

A major focus of this book, and in the media, has been on reducing carbon dioxide emissions. CO_2, after all, lasts thousands of years in the atmosphere, in prodigious amounts. But methane, in the short term, poses a much bigger threat: it has more than eighty times the climate warming impact of CO_2 over the first twenty years of its release.

The world's best hope to slow climate change in the near term is to cut methane quickly and dramatically, according to a May 2021

United Nations report. If cut by nearly half by 2030, a half-degree of warming could be prevented by mid-century, it noted.

> The report said the methane reduction would be relatively inexpensive and could be achieved by plugging leaks in pipelines, stopping venting of natural gas during energy drilling, capturing gas from landfills and reducing methane from belching livestock and other agricultural sources, which is the biggest challenge. Because methane helps make smog, cutting annual emissions of the gas by 45% or nearly 200 million tons could potentially prevent about 250,000 deaths a year worldwide from pollution- triggered health problems, the U.N. said (Ocko, et al., 2021).

However, the news on the methane front is not good. "... Study lead author Drew Shindell, a Duke University Earth sciences professor, said recent acceleration of methane pollution 'is really taking us far off' the Paris goals.'

Methane concentrations increased more in 2020 than since records began in 1983 and scientists don't know why. "After plateauing in the early 2000s, atmospheric methane has been increasing steadily since 2007. '2020s increase was double the 2007 growth. It's even higher than the early 1980s, when the gas industry was going crazy. It's really scary,' geophysicist Euan Nisbet of Royal Holloway, University of London told *New Scientist*'" (Vaughn, 2021).

To deal with the methane emergency, the U.S. and the European Union developed the Global Methane Pledge, in September 2021. Its goal: reduce global methane emissions by at least 30% by 2030 from 2020 levels. By mid-November, more than 100 countries had signed the pledge. However, the top three emitters, China, Russia and India had not.

If the pledge is successful, it could reduce warming by at least 0.2°C (0.36°F) by 2050, according to the White House. Methane, it's said, has been responsible for about one-half of the 1.1° C (1.98°F) warming that's occurred since the 1850s (White House, 2021).

A Coming Attraction

With worldwide air conditioning demand expected to massively increase this century—one study predicts as much as thirty-three-fold by 2100—solutions to reduce this are vitally important (Henley, 2015). Engineers at Purdue University have invented a unique solution: the whitest paint possible, which could cool buildings significantly enough to reduce, or even eliminate, the need for air conditioning.

In August 2021 the team developed that paint, to keep surfaces cool. "If you were to use this paint to cover a roof area of about 1,000 square feet, we estimate that you could get a cooling power of 10 kilowatts. That's more powerful than the central air conditioners used by most houses," said Xiulin Ruan, Purdue professor of mechanical engineering. Figure 21 shows Professor Ruan holding a sample of the paint.

The paint reflects up to 98.1% of sunlight and "sends infrared heat away from a surface at the same time." The paint's extreme whiteness comes from a "very high concentration" of barium sulfate, "which is also used to make photo paper and cosmetics white." Different sizes of barium sulfate are used. "How much each particle scatters light depends on its size, so a wider range of particle sizes allows the paint to scatter more of the light spectrum from the sun" (Wiles, 2021).

Tests showed the paint keeps surfaces nineteen degrees F. cooler than ambient surroundings at night. "It can also cool surfaces

8°F. below their surroundings under strong sunlight during noon hours" (Wiles, K., 2021). The paint is not yet available to purchase; it's undergoing certification testing.

Figure 21
Professor Ruan holding sample of whitest paint ever;
may significantly reduce air conditioning costs

Another option is passive house design. Ventilation systems naturally driven by heat can be integrated into new and existing buildings to keep them several degrees cooler than the ambient temperature (*Earthweek*, November 5, 2023).

CHAPTER 14

A Few Closing Words

> "The world will not be destroyed by those who do evil, but by those who watch them without doing anything."
> Albert Einstein

> "What do you need to solve the climate crisis? The answer is, everyone."
> Katharine Hayhoe, climate scientist at Texas Tech Univ., chief scientist for The Nature Conservancy, author of *Saving Us*

AS A COUNSELOR, I think in terms of diagnosis, symptoms, treatment and prognosis. What's our home's climate change diagnosis? Chronic, severe global warming disorder. Symptoms include the loss of the experience of home (aka solastalgia), as evidenced by hotter air temperatures, recurring wildfires and wildfire smoke, yearly toxic algae blooms, lower lake levels and late-season river and stream flows. In addition, diminished and earlier melting snowfall and less consistent and thinner winter lake ice.

Etiology (i.e., cause) of our global warming disorder: human activity, specifically burning fossil fuels (coal, oil and natural gas) and deforestation.

Treatment: primarily transition to clean, renewable energy systems as quickly as possible. Major barriers to successful treatment: personal, corporate and governmental denial, misinformation and willful ignorance driven by short-term profit (greed), addiction to fossil fuels, unbridled capitalism and belief systems not grounded in reality.

Prognosis: Guarded to poor, given the crisis situation and the short time for major actions. Successful treatment is dependent on acceptance of the disorder and its human cause, a willingness to make major changes and to tolerate current and future short-term distress and home loss for long-term climate stability and restoration of the home environment.

Severe global warming disorder, then, is far worse than its treatment. Indeed, successful treatment will improve physical and mental health, air quality, quality of life, reduce traffic congestion, restore a sense of home as well as save individuals and communities money.

As with most disorders, there are better and bad days, and in the case of climate disruption, better and bad years. But over time, without the necessary actions taken, the condition worsens, especially with such a chronic and severe "illness." The process of losing home intensifies.

A few words about denial and willful ignorance. Both your doctor, trained on the body's condition, and climate scientists on the climate system, apply the scientific method to their areas of expertise. Weather forecasters do as well. Therefore, denying human-caused climate change would be akin to telling your doctor, the weather forecaster and climate scientists alike that they're wrong about their areas of expertise.

Personally, there was no way our family was going to risk my brother's life, more than twenty years ago, on anything other than

A FEW CLOSING WORDS

medical science. Today, we can't risk inaction in the face of climate chaos to people's belief systems. There is too much at stake, and the results of science-based actions too convincing, to do so.

Now about willful ignorance and how it relates to climate disruption. Willful ignorance is an ideology, or what the late emeritus professor of religion at New York University James P. Carse, in *The Religious Case Against Belief,* calls a belief system. He provides the following example, one of many, in his book.

Galileo saw the moons of Jupiter and invited his skeptical colleagues on the faculty of the University of Padua to look through his telescope to see for themselves. Some refused outright, knowing that their Aristotelian understanding of the world (that Earth is the center of the universe) was at stake. Some looked, but incredibly reported they saw no moons.

[Yet] "The phenomena he [Galileo] described were already in plain view. If they were not noticed, it can only be that the viewer decided not to notice them, even while looking directly at them. How else to describe this, but as an act of *willful* ignorance" (Carse, 19).

Those of us living here for some time see changes in the climate (e.g., it's hotter and there's less snow than before). However, the willfully ignorant, or those bound to a belief system, do not, will not, or cannot see what is there and why it's happening. Due to religious, political, unresolved trauma, or other factors, they are blind to the truth: that human-driven climate disruption is the only viable explanation for what we are witnessing here, and around the world.

Since I began working on this book nearly six years ago, its emphasis has shifted significantly. Initially, it was a cautionary tale about global warming happening here. Now, there's an urgency, understood by many more, to protect one's home, here and around the world. And most recently, to use artificial intelligence in that pursuit.

During November's COP28, the United Nations climate summit in Dubai, officials from Google and Boston Consulting Group predicted that AI could help mitigate as much as one-tenth of all greenhouse gas emissions by 2030. Also, AI might be able to "speed up the recovery and design of new materials for low-emission energy technologies like advanced batteries."

In addition, businesses said AI is helping them "deliver alerts to people at risk of experiencing flooding, send text messages with hyperlocal planting advice to farmers coping with drought and help people exposed to high levels of air pollution decide the safest times to venture outdoors."

However, in AI's ability to produce insights and efficiencies that far exceed what today's computers can do, it devours enormous amounts of energy. "A peer-reviewed analysis in October estimated that AI systems worldwide could use as much energy in 2027 as all of Sweden." That electricity consumption could send emissions soaring and make global warming worse. Then there's the fear AI could someday pose a risk of extinction to humanity (Tankersley, *New York Times*, 2023).

Worldwide, 2023 was a scary summer. First, June was the hottest on record, then July, and then August. Eventually, the entire year not only broke, but shattered, temperature records. Deeply unsettling news, with no respite in sight.

A recent study noted that the record heat on land and in the water (101°F off the Florida coast!) around the Northern Hemisphere during the 2023 summer would be "virtually impossible" if climate change wasn't occurring. Another study warned that the vital conveyor belt of ocean currents in the Atlantic which controls climate could collapse by mid-century—or possibly any time after 2025—due to human-driven climate change (Ditlevsen & Ditlevsen, 2023).

We've lost more of our home's gifts of nature in the past five years. Thankfully, more people are aware of the threat and dedicated to preserving the habitability of our home. In the grandest sense, that includes the Earth. Consider something called Project Drawdown.

Project Drawdown has researched 100 innovative solutions to worldwide global warming, all commonly available, economically viable, and scientifically valid. They're described in *Drawdown: The Most Comprehensive Plan Ever Proposed to Reverse Global Warming*, edited by Paul Hawken (2017).

One highly effective solution is educating girls. There doesn't seem to be an obvious connection. However, when girls complete twelve years of education, they have almost four to five fewer (and healthier) children than a woman with no years of schooling. They realize higher wages, have a lower mortality rate, and they're less likely to marry as children or against their will. Educated girls are more resilient to climate change impacts as well.

Drawdown continues:

> It is the most powerful lever available for breaking the cycle of intergenerational poverty while mitigating emissions by curbing population growth. A 2010 economic study shows that investment in educating girls is 'highly cost-competitive with almost all of the existing options for carbon emissions abatement'—perhaps just $10 per ton of carbon dioxide (Drawdown, p. 81).

Quoting a United Nations study, *Drawdown* says that achieving "universal education in low- and lower-middle-income countries

… could result in 59.6 gigatons of emissions reduced by 2050. *The return on that investment is incalculable*" (Winthrop and Karas, 2015; cited in *Drawdown*, p. 82).

Then there are personal actions we can take, including supporting Our Children's Trust in their national and worldwide support of children's right to life, liberty and a stable climate. That can be direct financial support to the non-profit or sending emails to President Biden or the U.S. Dept. of Justice.

In the emails I've sent, I pointed out the hypocrisy of the administration publicly professing the seriousness of climate change (calling it "an existential threat") yet supporting the continued burning of fossil fuels and leasing of federal lands for coal, oil and natural gas extraction. But it was worse than that, I recently learned, although with a happier ending.

The cumulative CO_2 pollution threatened by our country's planned fracking of oil and gas and exports of liquefied natural gas (LNG) at more than twenty new LNG export facilities could add the equivalent emissions of 675 coal-fired power plants. The facilities would be centered in the Permian Basin of Texas and New Mexico and along the U.S. Gulf Coast (Oil Change International, 2023).

If these plans had come to fruition, we would have lost our home for a very long time, with ever worsening consequences for our children, grandchildren and beyond. Therefore, along with thousands of others, I sent a blistering email to the President and Department of Energy Secretary Granholm. The result, in late January 2024, was a "huge victory," in the words of Bill McKibben.

On January 26th the Biden Administration halted all new licenses for LNG export terminals, "until the policies used to figure out if they're in the 'public interest' can be updated to include modern economics and science" (McKibben, 2024). McKibben was thrilled at the news: "This is the biggest check any president has ever applied to the fossil fuel industry and the strongest move against dirty energy in American history" (McKibben, 2024).

Closer to home, talk about climate change whenever you get the chance. Discuss it with others. Find out what matters to them by asking and listening, then speak to how climate change impacts it, as well as its practical, smart solutions. Trust, as well as authentic engagement, flow from such active listening.

Share with others what you appreciate, or love, about living here. What makes it home for you? What outdoor activities, or seasons, do you most enjoy? Share the beauty of our local lakes, walking or riding around town on a bike, playing tennis. Maybe for you it's enjoying a picnic and/or music in the park, skiing in our nearby mountains or visiting family.

Support the electrification of as much of our economy as possible, including purchasing an electric vehicle when the time comes. If possible, use solar power to run the chargers, either with your own panels, or pay to use Avista's solar energy, through their My Clean Energy program. If not, be grateful that clean hydroelectric dams provide half of our power needs here (not without serious impacts, though).

If you don't have electric heat, minimize your natural gas use, both to reduce methane leaks along the extraction, distribution and use pipeline, and recently discovered toxic benzene exposure in your home. Look into an induction cook stove or stove top. If you can, convert to a cold-weather heat pump, with tax incentives. Maybe the window model, made by Gradient, when it's available. You can install it yourself.

Minimize your flying. Take Amtrak or drive to visit family, or vacation, across the country. Stay longer in one place during overseas vacations, and take far fewer of them. Work as much as possible online, to limit driving and flying.

The Love of Home

The following is a powerful story by Nigerian poet and novelist, Ben Okri, set 20,000 years in the future. It speaks to the "invisible" cost of our day-to-day actions and unseen systems.

> And Peace Shall Return
>
> They fell into the 'invisible effects' fallacy. This holds that though we do things which add up to a catastrophe, we are unable from moment to moment to see the effect of what we are doing. And because we don't see the effect from moment to moment, we therefore think that there isn't an effect. This fallacy enables people to continue their suicidal relationship with the earth right up to the very last moment of its devastation. More than any other fact, this is perhaps the most startling....Maybe the final suicide of humanity was not really some momentous event that happened in the last days, the drowning of the cities and the poisoning of the air and the detonation of nuclear bombs, but was caused by the mesmerism of a people chained to their past, the fatalism of a race that had accepted for centuries that it was simply unable to change. In this at least they were consistent. They perished, little by little, every day. They were killing themselves, and it didn't seem to trouble them.

To avoid perishing little by little, cherish what you appreciate, or love, about living here. One way to do this, paradoxically, is to give voice to what's been lost. Doing so, and giving space to experience and express the resulting grief, opens the heart to action.

We don't often realize that loss, not fully expressed, keeps us stuck in the past, unable to act.

Grief fully expressed, though, moves the body's stuck energy.

For example, what's it like for you to feel the stifling heat of *three consecutive years* of unprecedented heat waves? Have you felt trapped indoors for days, or a week or more, unable to see the beauty of our home, due to wildfire smoke? Have you become depressed or anxious? Have you experienced fear at not being able to breathe during once glorious summer days?

What's it like to:

- Be unable to ice fish where you did before?
- Worry about the ice being too thin when you ice skate or ride a snowmobile?
- Have less snow for kids to build snowmen with, or sled down Cherry Hill on?
- Have less, or poorer quality snow, to ski on?
- Have lower river flows when fishing late in the season?

What do those losses feel like? You may feel grief of what had been the essence of the home you've known being lost. Name the emotion, if you can, and feel it, until it completes. If there's frustration or anger, allow it to move you to appropriate action, whatever that is for you.

What else can a person do? Live simply. Don't waste time or money chasing the latest gadget of distraction. Instead of driving, ride a bike, walk or take public transit whenever possible, eat less meat. Dry washed clothes outside whenever possible, and let clean dishes air dry overnight. Run the dishwasher, and the clothes washer, only when they're full. Compost food scraps and other organic material. Limit, or consciously monitor, your engagement with social media.

There's no recipe for living simply. What works for me won't necessarily work for you. If you're interested, I recommend Duane Elgin's 2010 book, *Voluntary Simplicity*, for many great ideas, as well as Graham Hill's inspiring 2013 article, "Living with Less. A Lot Less."

Ginny and I recycle and compost. We use our own bags when we go to Costco. I eat what I take, to not waste food. Unfortunately, so much food is wasted when I volunteer once a week at a local middle school.

We keep the heat no higher than 67°F. in the winter. Using a blanket, long-sleeved flannel shirts and slippers keeps me plenty warm. I take showers only when I need one and turn off the shower when I soap up, using hot water only to rinse off. We shovel the snow off our sixty-foot driveway together. We appreciate burning off extra calories.

While our habits are not for everyone, most people can find ways to live simpler, more fulfilling lives. We understand that our low-impact lifestyle won't stop global heating, but neither of us live it for that reason. We just appreciate living simply and staying active. Living that life, while engaged in worthwhile activities to stem the threat, is the antidote to despair.

Why else to live more simply? It improves the quality of our relationships with people we love. Gluttony, unconscious, distraction-driven behavior costs us our ability to be with those people we care about. They know what we value by where we spend our time and money.

If it's a bauble, obsessively watching TV, playing a video game or scrolling on social media for hours instead of being with people you love, they get the message that they matter less than that thing or behavior. That isolates them and you, perpetuating the very loss of connection that's at the heart of so many of our individual and societal issues.

Another idea, offered by this book's editor: own only what you need and use it. Constant buying and hoarding stuff without employing it in some way is wasteful and it clogs up the house and garage. How many candles and DVDs does one really need?

Everyone ought to have a basic understanding of climate change and what they can do to minimize and adapt to it. Then pass your life choices on to the next generation, as mine were by my parents and grandparents, even though they knew nothing about carbon pollution. Call it a Generational Oath, similar to the Native American commitment to act with seven future generations in mind.

Push local elected officials to incentivize less vehicle use and implement mixed-use, clustered developments, which provide nearby services we need and want. Walking or biking also enhances community engagement, by exposing us to other points of view, spurs the community's economic vitality as well as our own physical and mental health. Join me in enhancing the walkability of our towns.

Prioritize safety, not speed, on the roads. Don't support expanding highways to handle more traffic. This only induces greater vehicle use, and longer trips (called "induced demand"). Vote to fully fund transit services as well.

Join Others

Bill McKibben's recommendation to "become less of an individual" to make a difference with climate change brings up the question: What group or movement could I join?

LOSING HOME

Some local organizations:

- Kootenai Environmental Alliance
- Model Forest Policy Program (based in Sandpoint, ID)
- 350.org Sandpoint or Spokane
- Idaho Conservation League
- The local Citizens Climate Lobby affiliate
- Gonzaga's Center for Climate, Society, and the Environment

Nationally:

- 350.org
- Our Children's Trust
- Sunrise Movement
- Extinction Rebellion
- Fridays For Future
- Sierra Club
- Natural Resources Defense Council
- Environmental Defense Fund
- Earth Justice
- The Nature Conservancy.

There are other groups I'm sure I missed. The point, though, is that there are many people, around the world, and around the corner, engaged to protect our homes from the ravages of climate change. It's up to us, but not us alone, to protect our home, wherever that may be.

References

Abatzoglou, John T. and A. Park Williams. "Impact of anthropogenic climate change on wildfires across western US forests." *Proceedings of the National Academy of Sciences* 113, no. 42 (2016): 11770-11175.

Air Force Times. "Tyndall sustains direct hit, extensive damage from hurricane." October 11, 2010.

Albrecht, Glenn. "The age of solastalgia." *The Conversation* 7, 2012.

Allsup, Maeve and Lisa Martine Jenkins. "DOE moves to streamline transmission permitting." *Latitude Media*, April 25, 2024.

Ames, Daniel L. and Susan T. Fiske. "Intentional Harms Are Worse, Even When They're Not." *Psychological science* 24, no. 9 (2013): 1755-1762.

An Illustrated History of North Idaho. Western Historical Publishing Company (1903).

Anderegg, William RL., John T. Abatzoglou, Leander DL Anderegg, Leonard Bielory, Patrick L. Kinney, and Lewis Ziske. "Anthropogenic climate change is worsening North American pollen seasons." *Proceedings of the National Academy of Sciences* 118, no.7 (2021): 6e2013284118.

Apolo Heating & Air Conditioning. "Why Can't I Keep a Cool Home with the 100 Degree Heat Wave?" (2023).

"Area Still Staggering Under Heavy Snow and Winds." *Coeur d'Alene Press*, January 27, 1969.

Arenschield, Laura. "Toxic algae bloom now stretches 650 miles along Ohio river." *The Columbus Dispatch*, October 3, 2015.

Askland, Hedda Haugen, Barrie Shannon, Raymond Chiong, et al. "Beyond migration: a critical review of climate change induced displacement." *Environmental sociology* 8, no. 3 (2022): 262-278.

Asthma & Allergy Foundation of America. (2023). "2023 Allergy Capitals: Top 100 Most Challenging Cities to Live in With Allergies."

Asthma & Allergy Foundation of America. (2023). "Asthma Capitals: Top 100 Most Challenging Cities to Live in With Asthma."

Asthma & Allergy Foundation of America. Asthma Surveillance Data. CDC.gov. (2018).

Asthma & Allergy Foundation of America. CDC.gov. (2018). Asthma | Healthy Schools.

Avista Utilities. (2020). Wildfire Resiliency Plan. Identified within Avista's WUI area north of Coeur d'Alene is the town of Hayden Lake and further north, every town in Idaho east of Coeur d'Alene, including Cataldo, Pinehurst, Enaville, Prichard, Kellogg, Wallace, Mullan and Murray, on both sides of the I-90 freeway. South of Coeur d'Alene are towns including Harrison and St. Maries and other smaller towns.

Ball, Jeffery. "The Carbon Underground." *Wired*, Sept. 2022.

Barry, Keith. "Will an Electric Car Save You Money?" *Consumer Reports*, February 16, 2023.

Basagana, Xavier, Claudio Sartini, Jose Barrera-Gõmez, Payam Dadvand, Jordi Curillera, Bart Ostro, Jordi Sunyer, and Mercedes

Medina-Ramón. "Heat waves and cause-specific mortality at all ages." *Epidemiology* 22, no. 6 (2011): 765-772.

Baumhardt, Alex. "Feds plan to add NW, Western state land to potential expansion of solar energy." *Oregon Capital Chronicle.* Cited in *The Spokesman-Review,* January 21, 2024.

Belkin, Lisa. "'Barometer of despair': Birthrate falls as millennials fear climate apocalypse." *Yahoo!news,* November 6, 2019.

Berners-Lee, Mike. "How (Not) To Buy." Greta Thunberg., *The Climate Book: The Facts and the Solutions* (New York: Penguin Press, 2022), Part IV: chapter 20.

Berton, Elena. "Flight shaming hits air travel as 'Greta effect' takes off. *Reuters,* October 2, 2019.

Bittle, Jake. "Why cooling centers sit empty." *Grist,* August 22, 2023.

Blue, Elly. "The Free Rider Myth — Who Really Pays for the Roads," *Momentum Mag,* March 24, 2016. From Jeff Speck's *Walkable City Rules: 101 Steps To Making Better Places.* (Washington, D.C.: Island Press, 2018), 3.

Borud, Matthew. "Our Gem Speaker Series: Economy of our Ecology." Idaho Department of Commerce, March 30, 2022. https://www.uidaho.edu/cda/cwrc/our.gem/videos.

Bouchama, Abderrezak, Mohammed Dehbi, Gamal Mohamed, Franziska Matthies, Mohamed Shoukri, and Bettina Menne. "Prognostic Factors in Heat Wave-Related Deaths: A Meta- analysis." *Archives of internal medicine* 167, no. 20 (2007): 2170-2176.

Brand, Christian. "Cycling is ten times more important than electric cars for reaching net-zero cities." *The Conversation*, March 29, 2021.

Brocious, Ariana. "Nearly 2 Years In. Is the Inflation Reduction Act Delivering Yet?" Climate One podcast interview with Houser, April 12, 2024.

Brooks, David. "How to Know a Person. The Art of Seeing Others Deeply & Being Deeply Seen." *You Tube,* November 1, 2023.

Brown, Alex. "Feds have money to help save homes from wildfire. Now, can they get the word out?" *The Oregonian* (Portland, Oregon), August 28, 2023.

Buis, Alan. "The Atmosphere: Getting a Handle on Carbon Dioxide." *NASA's Jet Propulsion Laboratory,* October 9, 2019.

Buckley, Cara. "These Cities Aren't Banning Meat: They Just Want You to Eat More Plants." *New York Times*, February 28, 2024. Cited in billmckibben@substack.com, "Is $30 trillion a lot?" April 17, 2024.

Buley, Bill. "Unprecedented." *Coeur d'Alene Press*, July 1, 2021.

Bullard, Nathaniel. "EV sales continue growing at double digits, now only growing auto market." *Bloomberg*, in *Spokesman-Review* (Spokane, Washington), June 9, 2023.

Caiazzo, Fabio, Akshay Ashok, Ian A Waitz, Steven HL Yim, and Steven RH Barrett. "Air pollution and early deaths in the United States. Part 1: Quantifying the impact of major sectors in 2005." *Atmospheric Environment* 79 (2013):198-208. Report cited in Jeff Speck's *Walkable City Rules: 101 Steps To Making Better Places* (Island Press: Washington, D.C.: 2018).

Carbon Bankroll 2.0: From Awareness to Action report. https://www.topofinance.org/carbon-bankroll. Cited in billmckibben@substack.com, "Is $30 trillion a lot?" April 17, 2024.

Carbon Pricing Dashboard. *The World Bank,* https://carbonpricing-dashboard.worldbank.org.

Carlson, Debbie, "Engine No. 1 says it's new representation on ExxonMobil's board has already scored a win," MSN, *Money Watch*, September 24, 2021.

Carroll, Matthew and Travis Paveglio. "Using community archetypes to better understand differential community adaptation to wildfire risk." *Philosophical Transactions of the Royal Society B: Biological Sciences* 371, no. 1696 (2016):20150344.

Carse, James P. *The Religious Case Against Belief* (New York: Penguin Books, 2008).

Carter, Therese S., Colette L. Heath, and Noelle E. Selin. "Large mitigation potential of smoke $PM_{2.5}$ in the US from human-ignited fires." *Environmental Research Letters* 18, no. 1 (2023): *014002*.

CBS News. "The climate change lawsuit that could stop the U.S. Government from supporting fossil fuels." March 3, 2019. [Overtime] Retrieved Sept. 18, 2020.

CdA Tribe>History https://www.cdatribe-nsn.gov/our-tribe/history/. The Tribe's website states that it has been here "since time immemorial."

Center for Biological Diversity. "Global Warming and Endangered Species."

Centola, Damon. Change: How to Make Big Things Happen (London: John Murray Press, 2021).

Cerny, Elaine, "Iced tea, anyone?" *Coeur d'Alene Press*, July 19, 2015.

Chandler, David L. "Explaining the plummeting cost of solar power." *MIT News Office,* November 20, 2018.

Chiu, Allyson. "You could be doing laundry wrong." *The Washington Post.* Cited in *The Spokesman-Review,* January 28, 2024.

Chung, Connor & Dan Cohn. "Passive Investing in a Warming World. An Evaluation of Fossil Fuel Impacts on Equity Portfolios." *Institute For Energy Economics and Financial Analysis*, February 2024. Cited in "Fossils drag the market," Tim McDonnell, *Semafor Net Zero,* February 16, 2024.

Clark, Rachel. "Mount Spokane closes due to lack of snowfall." *The Gonzaga Bulletin,* February 19, 2015.

Clayton, Susan, Christie Manning, Kirra Krygsman, and Meighen Speiser. "Mental health and our changing climate: impacts, implications and guidance." (Washington DC: *American Psychological Association and ecoAmerica*, 2017), 22.

Climate Change on the Coeur d'Alene Landscape: Climate Impact Assessment for the Coeur d' Alene Reservation and Aboriginal Territory. *Coeur d'Alene Tribe, Natural Resources Department, Environmental Programs Office — Climate Division.* 2023.

"Closures Continue For Week." *Coeur d'Alene Press,* January 29, 1969.

Clouse, Thomas. "Avista makes deal to get out of aging Colstrip power plant." *Spokesman- Review* (Spokane, Washington), August 11, 2023.

Clouse, Thomas and Tod Stephens. "Forecast: Rolling Blackouts: Avista to Begin Cutting Power This Summer During Storms to Avoid Fires." *Spokesman-Review,* May 8, 2024.

"Coeur d'Alene River Ranger District Issues Decision on Honey Badger Project." *U.S. Forest Service,* Coeur d'Alene River Ranger District, Idaho, January 12, 2022.

Corbin, Clark. "Idaho receives $28 million to build EV charging stations every 50 miles of Interstate," *Idaho Capital Sun* (Boise, Idaho), July 12, 2022.

Costley, Drew. "Study calls for strict limits on oil, coal to curb warming." *Spokesman-Review,* September 14, 2021.

Cousins, Keith. "Everyone deserves a medal today." *Coeur d'Alene Press,* June 29, 2015.

Cousins, Keith. "Praying for rain." *Coeur d'Alene Press,* August 21, 2015.

Cuniff, Meghann M. "Snowfall puts CdA in record territory," *Spokesman-Review,* Feb. 6, 2008.

REFERENCES

Dahlgren, Dorothy and Simone Carbonneau Kincaid. *In All the West No Place Like This: A Pictorial History of the Coeur d'Alene Region, Revised.* (Coeur d'Alene, Idaho, Museum of North Idaho, 1996).

Dalton, Greg and Ariana Brocious. "Just a Walk or Bike Ride Away: The 15-Minute City." Climate One podcast. August 11, 2023. Accessed August 14, 2023. Reduce unneeded parking as well (e.g., surrounding malls and shopping centers) to reduce heat-island effects, which dramatically raise nearby temperatures, as well as PM 2.5, PM 10 and greenhouse gas emissions. See, for example, two *Sightline Institute* articles by Catie Gould, Dec. 16, 2021 and January 11, 2022.

"Dangerous ground." *Coeur d'Alene Press,* January 13, 2021.

Davenport, Coral. "White House may slow early stage of shift to electric cars." *The New York Times.* Cited in the *Spokesman-Review,* February 18, 2024.

Davies, Rob. "Toyota claims battery breakthrough, in potential boost for electric cars." *The Guardian,* July 4, 2023.

Davis, Steve. 'Bigger vehicles are directly resulting in more deaths of people walking." *Smart Growth America,* April 12, 2021.

Declaration of Dr. Van Susteren, para. 12.

Declaration of Lise Van Susteren in Support of Plaintiffs' Urgent Motion under Circuit Rule 27-3(b) for Preliminary Injunction, paras. 13, 28. Cited in Speth, James Gustave. *They Knew: The US Federal Government's Fifty-Year Role in Causing the Climate Crisis.* Cambridge, MA: The MIT Press, 2021, 142.

Department of Defense. Report on Effects of a Changing Climate to the Department of Defense. January 10, 2019.

DEQ website. https://www.deq.idaho.gov/water-quality/surface-water/cyanobacteria-harmful- algal-blooms/.

Deshais, Nicholas. "Urban thinker sizes up how we move." *Spokesman-Review* (Spokane, Washington), July 23, 2018.

Dessler, Andrew. "Is nuclear energy the answer?" www.theclimatelink.com/p/is-nuclear-energy-the-answer?. Cited in billmckibben@substack.com, "Is $30 trillion a lot?", April 17, 2024.

DeVoe, Jo. "'Fifteen-minute cities,' gaining traction in a post-pandemic world, are already here in Arlington." *ARLnow*, December 20, 2022. https://www.arlnow.com/2022/12/20/fifteen-minute-cities-gaining-traction-in-a-post-pandemic-world-are-already-here-in-arlington/.

Dias, Elizabeth. "Pope Francis Urges Climate-Change Action in Encyclical." *Time*, July 23, 2018.

Ditlevsen, Peter and Susanne Ditlevsen. "Warning of a forthcoming collapse of the Atlantic meridional overturning circulation." *Nature Communications* 14, no. 1 (2023): 1-12. https://doi.org/10.1038/s41467-023-39810-w.

Dodgen, D., D. Donato, N. Kelly, A LaGreca, J. Morganstein, J. Reser, J. Ruzek et al. "The impacts of climate change on human health in the United States: A scientific assessment." *US Global Change Research Program* (2016): 217-246.

"Downtown Streets Now Under Floods: Predict Frigid Weather for Flood District." *Spokesman-Review* (Spokane, Washington), December 25, 1933.

Dreher, Arielle. "Heat wave was a 1 in 1,000-year event made more likely by climate change, study shows." *Spokesman-Review* (Spokane, Washington), July 8, 2021.

Duff, Mike. "Toyota Lays Out Its EV Battery RoadMap, Including a Solid-State Battery (Eventually)." *Car and Driver.* November 26, 2023.

Dumont, Caroline, Elizabeth Haase, Trygve Dolder, Janet Lewis, and John Coverdale. "Climate change and risk of completed suicide." *The journal of nervous and mental disease* 208, no. 7 (2020): 559-565. Mortality by State, as of March 1, 2022. *Mortality & Morbidity Weekly Report* 59 (1009). Idaho's death rate per 100,000 population is 23.2. Wyoming has the highest suicide death rate, at 30.5; Alaska is second at 27.5; Montana is #3 at 26.1. Washington's rate is much lower, #26, at 17.1 per 100,000.

Earth Observatory, NASA. World of Change: Ice Loss in Glacier National Park. earthobservatory.nasa.gov/world-of-change-ice-loss-in-glacier-national-park.

Earthweek. (July 30, 2023). Distributed by Andrews McMeel Syndication. Cited in *The Spokesman-Review.*

Earthweek. (November 5, 2023). Distributed by Andrews McMeel Syndication. Cited in *The Spokesman-Review.*

Edelen, Amy. "Wall Street Journal/Realtor.com names Coeur d'Alene top emerging housing market in U.S.; Spokane No. 5." *Spokesman-Review* (Spokane, Washington), April 28, 2021.

"EIA's Open Data API, Electricity Net Generation." *U.S. Energy Information Administration*, 2022.

"Electricity Data Browser, Net generation, United States, all Sectors." *U.S. Energy Information Administration.* https://www.eia.gov/electricity/data/browser/.

Ellersick, Loreen. Obituary. 1923-2017. https://www.flammfh.com/obituary/4241042.

Emerson, Tom. *Fred Murphy: A Legend of Coeur d'Alene Lake.* (Coeur d'Alene, ID: Century Publishing, 1988).

Energy Information Administration. "What is U.S. electricity generation by energy source?" February 29, 2024.

Environmental Defense Fund. "Nine ways we know humans cause climate change."

EPA rule. "Biden-Harris Administration Finalizes Standards to Slash Methane Pollution, Combat Climate Change, Protect Health, and Bolster American Innovation. U.S. Environmental Protection Agency, December 2, 2023. Cited in Johnson, Jonathan. "Is Biden waging a war on energy? Or one on climate?" *High Country News*, December 30, 2023.

Erickson, Keith. "Storm bill may hit $16 million." *Coeur d'Alene Press*, December 6, 1996.

Expert witness testimony of Dr. Daniel Fagre, *Held v State of Montana*, CDV-2020-307, June 13, 2023.

Expert witness testimony of Dr. Lori Byron, *Held v State of Montana*, June 13, 2023.

Expert witness testimony, Mark Jacobson, *Held v. State of Montana*, June 16, 2023.

Federal Register, 2015.

Filazolla, Alessandra, Kevin Blasgrave, Mohammad Arshad Imrit, and Sapna Sharma. "Climate change drives increases in extreme events for lake ice in the Northern Hemisphere." *Geophysical Research Letters* 47, no. 18 (2020): e2020GLO89608.

Findings of Fact and Conclusions of Law. *Held v. State of Montana*, July 15, 2023, 15. Judge Kathy Seeley of Lewis & Clark District Court, Helena, Montana.

"Fires Prompt Temporary Closures in the Idaho Panhandle National Forest." *Coeur d'Alene Press*, July 29, 2021.

Folkman, Susan, Margaret Chesney, Leon McKusick, Gail Ironson, David S. Johnson, and Thomas J. Coates. "Translating coping theory into an intervention." *The social context of coping*. (1991): 239-260.

REFERENCES

"Ford Foundation nixes fossil fuels." *Spokesman-Review,* October 19, 2021.

"Ford to delay rollout of new electric pickup and SUV as EV sales slow." *The Guardian,* April 4, 2024.

FOTW #1230, "More Than Half of all Daily Trips Were Less than Three Miles," March 21, 2022. https://www.energy.gov/eere/vehicles.

Fourth National Climate Assessment. Chapter 24: Northwest. KM1: Natural Resource Economy. *U.S. Global Change Research Program,* 2018.

Friedlingstein, Pierre, Matthew W. Jones, Michael O'Sullivan, Robbie M. Andrew, et al. "Global Carbon Budget 2021." *Earth System Science Data,* 14, no. 3, 1917-2005.

Future. "How much carbon-and money-you can save by biking." May 20, 2022.

Garnett, Matthew and Sally C. Curtin. Suicide Mortality in the United States, 2001-2021. NCHS Data Brief, no. 464, National Center for Health Statistics, 2023, Centers for Disease Control and Prevention.

Gates, Bill. *How to Avoid a Climate Disaster: The Solutions We Have and The Breakthroughs We Need.* (New York: Alfred A. Knopf, 2021).

Gates, Dominic. "Sustainable Aviation Fuel Startup Breaks Ground on Plant." *Seattle Times.* Cited in *Spokesman-Review,* July 13, 2023.

Gates, Dominic. "E-fuel firm to locate plant in Moses Lake." *Seattle Times.* Cited in *Spokesman-Review,* June 20, 2023.

Gerdes, Justin on X:".@ENERGY: https://twitter.com/JustinGerdes/status/1747331947377107139.

Gervais, Jennifer A., Ryan Kovach, Adam Sepulveda, Robert Al-Chokhachy, J. Joseph Giersch, and Clint C. Muhfield. "Climate-induced expansions of non-native species in the Pacific Northwest, North America: a synthesis of observations and projections." *Biological Invasions* 22, no. 7 (2020): 2163-2183. https://doi.org/10.1007/510530-020-02244-2.

Gilchrist, J., T. Haileyesus, M. Murphy, C. Collins, and N. McIlvain. *Heat Illness Among High School Athletes — United States, 2005-2009.* Morbidity & Mortality Weekly Report 59, no. 32 (2010):1009-1013.

"Global Catastrophe Recap: First Half of 2023." *Aon.* Updated on August 17, 2023.

"Global Catastrophe Recap Sept. 2021 (PDF) (Report). *Aon Benfield.* Oct. 12, 2021.

"Global Energy Demand for Fossil Fuels." *The Guardian,* September 11, 2018.

Gogage, Jacob. "Home insurers cut natural disasters from policies as climate risks grow." *Washington Post* (Washington, D.C.), September 23, 2023.

Goggin, Michael, Rob Gramlich, and Michael Skelly. "Transmission Projects Ready to Go: Plugging into America's Untapped Renewable Resources." Americans for a Clean Energy Grid, April 2021.

Goodkind, Andrew L., Christopher W. Tessum, Jay S. Coggins, Jason D. Hill, and Julian P. Marshall. "Fine-scale damages of particulate matter air pollution reveal opportunities for location-specific mitigation of emissions." *Proceedings of the National Academy of Sciences* 116, no. 118 (2019): 8775-8780.

REFERENCES

Governor's Office of Planning and Research, State of California, "List of Worldwide Scientific Organizations." https://www.opr.ca.gov/facts/list-of-scientific-organizations.html.

Graziano, Marcello and Kenneth Gillingham. "Spatial patterns of solar photovoltaic system adoption: the influence of neighbors and the built environment. *Journal of Economic Geography 15*, no. 4 (2015): 815-839.

Greene, Brenna. "Ironman competitors brace for hot temperatures on Sunday." KREM2, June 24, 2021.

Gruver, Mead. "Is safer nuclear power on the way?" Associated Press. Cited in *Spokesman- Review*, June 3, 2021.

Guy, Kate and Erin Sikorsky. "Blueprint for the Department of Defense's Strategic Assessment of Climate Change." Commentary. *War On the Rocks*, March 12, 2021.

Hagengruber, James. "Ice age returns to Hayden Lake," *Spokesman-Review*, January 17, 2007.

Hall, Steve M., Energy Report. *Idaho Climate-Economic Impacts Assessment.* James A. & Louise A. McClure Center for Public Policy Research, University of Idaho. Boise, ID., 2019.

Hallisey, Karen. "How Riding a Bike Benefits the Environment." *UCLA,* May 11, 2022.

Hanlon, James. "Faith and Climate Change." *Spokesman-Review*, March 10, 2023.

Hanna, Ryan. Interview with author, June 24, 2022.

Harris, Cliff. "Our updated city-by-city winter of 2014-2015 snowfall predictions." *Coeur d'Alene Press*, December 22, 2014.

Harris, Cliff. "A brief North Idaho weather review." *Coeur d'Alene Press*, December 29, 2014.

Harris, Cliff. "I'm still predicting all-time record snows by 2019-20." *Coeur d'Alene Press,* January 12, 2015.

Harris, Cliff. "The powerful El Nino of 1997-98 caused huge problems on a global scale." *Coeur d'Alene Press*, July 20, 2015.

Harris, Cliff. "2015 was a sleighride of wild extremes." *Coeur d'Alene Press*, January 11, 2016.

Hartmann, Thom. *The Last Hours of Ancient Sunlight: The Fate of the World and What We Can Do Before It's Too Late.* New York: Three Rivers Press, 2004.

Hausfather, Zeke. "Analysis: Why children must emit eight times less CO_2 than their grandparents. *Carbon Brief,* April 10, 2019, https://www.carbonbrief.org/analysis-why-children-must-emit-eight-times-less-co2-than-their-grandparents.

Hawken, Paul, ed., *Drawdown: The Most Comprehensive Plan Ever Proposed to Reverse Global Warming.* New York: Penguin Books, 2017.

Heilmann, Killian, Matthew E. Kahn, and Cheng Keat Tang. "The urban crime and heat gradient in high and low poverty areas." *Journal of Public Economics* 197 (2021): 104408.

Heller, Martin, Gregory Keoleian, and Diego Rose. "Implications of Future US Diet Scenarios on Greenhouse Gas Emissions." *Center for Sustainable Systems Report* (2020). Cited in Sandstrom, Summer, "Meatless Mondays Matter." *The Inlander*, April 20, 2023.

Henley, Jon. "World Set to Use More Energy for Cooling Than Heating." *The Guardian,* October 26, 2015. 26.

Henning, Brian G. *Riders in The Storm: Ethics in an Age of Climate Change.* (Winona, MN: Anselm Academic, 2015). Analogy is from Prof. Henning's students in his 2013 Ethics of Climate Change course at Gonzaga University.

Hickman, Caroline, Elizabeth Marks, Panu Pinkala, Susan Clayton, R Eric Lewandowski, Eloise E. Mayall, Britt Wray, Catriona

Mellor, and Lise Van Susteren. "Climate anxiety in children and young people and their beliefs about government responses to climate change: a global survey." *The Lancet Planetary Health* 5, no. 12 (2021): e863-e873.

Hijazi, Jennifer. "Montana Youth Win Historic Case on Harm from Climate Change." *Bloomberg Law,* August 14, 2023.

Horn, Gerhard. "Toyota Ready to Dominate the EV Market After Playing the Long Game." *CarBuzz.* July 9, 2022. https://www.carbuzz.com/news/toyota-ready-to-dominate-the-EV-market-after-playing-the-long-game/.

Houghton, Katherine. "Smoke Leaves Lung Damage Long After Air Clears." *Spokesman-Review,* September 20, 2020.

"How we can combat climate change. Opinions by Post Opinions Staff." *Washington Post,* January 2, 2019. In January 2023 Avista announced that it would end its partial ownership in the Colstrip coal-fired power plant on Dec. 31, 2025. Clouse, Thomas. "Avista makes deal to get out of aging Colstrip power plant in Montana," *Spokesman-Review,* January 22, 2023.

Idaho Dept. of Fish & Game. *Idaho State Wildlife Action Plan 2022,* August 2, 2022.

Idaho State Parks & Recreation study, 2017. Oct. 15, 2021 email from Josh Wise, Economic Development Specialist, Coeur d'Alene Area, Economic Development Council — Jobs Plus.

International Energy Agency. *CO2 Emissions in 2022,* March 2023.

Inglis, Bob. "Unmasking the deceit over climate change." *Greenville Journal* op-ed, February 3, 2017.

Introduction to Kootenai County. https://www.kcgov.us/586/introduction-to-kootenai-county/. Average yearly snowfall is 69.8 inches.

Ivanova, Diana, John Barrett, Dominik Waldenhofer, Biljana Macura, Max Callaghan, and Felix Creutzig. "Quantifying the potential for climate change mitigation of consumption options." *Environmental Research Letters* 15, no. 9 (2020): 093001.

Ivanova, Diana, Konstantin Stadler, Kjartan Steen-Olsen, Richard Wood, Gibran Vita, Arnold Tukker, and Edgar G. Hertwich. "Environmental impact assessment of household consumption." *Journal of Industrial Ecology* 20 no. 3 (2016): 526-536.

Jarvis, Andy, director of the Future of Food program at the Bezos Earth Fund. "One Good Text." Cited in *Semafor Net Zero*, January 24, 2024.

Johnson, Elena. "What lurks beneath the lake." *Coeur d'Alene Press*, June 5, 2019.

Johnson, Nathaniel. "GOP Rep. Mike Simpson: It's my party, and I'll fight climate change if I want to." *The Grist*, May 1, 2019.

"Joint US-EU Press Release on the Global Methane Pledge." *White House*, September 18, 2021.

Katz, Lily. "How Much Does Walkability Increase the Value of a Home?" *Redfin*, February 11, 2020.

Kootenai Electric Cooperative Annual Report, 2012.

Kolden, Crystal A. and Carol Henson. "A Socio-Ecological Approach to Mitigating Wildfire Vulnerability in the Wildland-Urban Interface: A Case Study from the 2017 Thomas Fire." *Fire* 2, no. 1 (2019): 9.

Kootenai County floodplain letter, June 2023. Distributed to county residents "very likely living in the floodplain."

Kootenai Power Cooperative *Power Lines* newsletter, May 2023.

Kramer, Becky. "Pacific Northwest's 2015 weather likely to be repeated, climate scientists say." *Spokesman-Review*, November 4,

2015. These projections were made at the 6th Annual Northwest Climate Change Conference in Coeur d'Alene.

Krauss, Clifford. "As Oil Companies Stay Lean, Workers Move to Renewable Energy." *New York Times.* Cited in *Spokesman-Review,* March 3, 2023.

KREM-2 News at noon, November 29, 2023.

Krisher, Tom and David McHugh, D. "Ford to go all electric in Europe by 2030." Associated Press. Cited in *Spokesman-Review,* February 18, 2021.

Krisher, Tom. "GM, Ford to outdo each other with electric vehicle investments." Associated Press. Cited in *Spokesman-Review,* June 17, 2021.

Kroman, David. "Washington state will ban new gas-powered cars by 2035, following California's lead." *Seattle Times.* Cited in *Spokesman-Review,* August 25, 2022.

L., Jennifer. "Toyota Reveals Solid-State Battery with 745-Mile Range, Cuts Emissions by 39%." *Carbon Credits.com,* July 5, 2023.

LaCroix, Trea. *A Nutrient Mass Balance of Fernan Lake, Idaho and Directions for Future Research.* University of Idaho, 2015.

Laczko, Leslie. "National and local attachments in a changing world system: Evidence from an international survey. *International Review of Sociology—Revue Internationale de Sociologie* 15, no. 3 (2005): *517-528.* Survey was the International Social Service Programme [ISSP], 1995.

Lagoo, Jeremy. "Spokane with the worst air quality in the United States, hazardous conditions remain." *KREM-TV,* August 23, 2019.

Laing, Keith. "Tesla to put EV chargers at 20,000 Hilton hotels." *Bloomberg.* Cited in *Spokesman-Review,* September 8, 2023.

Lambert, Jonathan. "Heat waves aren't just physically harmful — research shows they can harm mental health, too." *The Messenger*, July 22, 2022.

Laumatia, Laura, Environmental programs manager, Coeur d'Alene Tribe. Data compiled by Diana Rice, Data Analyst, Marimn Health.

Laumatia email, May 1, 2024.

Leber, Rebecca. "The air we breathe was getting better. Then climate change hit." *Vox*, April 19, 2023.

Legacy Industry Snapshot 10/21 — Tourism Industry. Oct. 2021 email from Josh Wise, Economic Development Specialist, CdA Economic Development Corp./Jobs Plus, Inc.

Leiserowitz, Anthony, Edward Maibach, Seth Rosenthal, John Kotcher, Emily Goddard, Jennifer Carman, Marija Verner, Matthew Ballew, Jennifer Marlon, Sanguk Lee, Teresa Myers, Matthew Goldberg, Nicholas Badullovich, and Kathryn Thier. "Global Warming's Six Americas, Fall 2023." *Yale University and George Mason University*. New Haven, CT: Yale Program on Climate Change Communication.

Lind, Treva. "Forces of nature: Doctors link environment and climate change to health issues." *Spokesman-Review*, August 7, 2019.

Linden, Eugene (2022). "The True Cost of Climate Change," Greta Thunberg, ed., *The Climate Book: Facts and Solutions* (New York: Penguin Press, 2022). Part III: Chapter 20.

Lines, William J. *Open air: Essays*. (Sydney, Australia: New Holland Publishers, 2001).

Loyd, Nic and Linda Weiford. "Triple-digit heat topples records across Inland Northwest once again." *Spokesman-Review*, August 17, 2023.

Loyd, Nic and Linda Weiford, L. "We had our 5th-hottest summer." *Spokesman-Review*, September 7, 2023.

Lutey, Tom. "Judge sides with NorthWestern in Laurel power plant suit." *Billings Gazette*, February 8, 2024.

Lutz, Catherine and Anne Lutz Fernandez. *Carjacked: The Culture of the Automobile and Its Effect on Our Lives.* (New York: St. Martin's Press, 2010). Cited in *Walkable City Rules: 101 Steps To Making Better Places.*

Lynn, E. A., Cuthbertson, M. He, J.P. Vasquez, M. Anderson, J.T. Abatzoglou, and B.J. Hatchett. "Technical note: Precipitation-phase partitioning at landscape scales to regional scales." *Hydrology and Earth System Sciences.* https://doi.org/10.5194/hess-24-2020.

Maibach, Edward, Jeni Miller, Fiona Armstrong, Omnia El Omrani, Ying Zhang, Nicky Philpitt, Sue Atkinson, et al. "Health Professionals, the Paris agreement, and the fierce urgency of now." *The Journal of Climate Change and Health* 1 (2021): 100002.

Mann, Randy. "The 'Dust Bowl' Years of the 1930s Featured Prolonged Droughts." *Coeur d' Alene Press*, September 3, 2012.

Mann, Randy. "2015 was a sleighride of wild extremes." *Coeur d'Alene Press,* January 11, 2016.

Mann, Randy. 'Slow-moving weather patterns bringing extreme weather conditions." *Coeur d' Alene Press*, September 4, 2017.

Mann, Randy. "Mother Nature leaves a pile of bills." *Coeur d'Alene Press*, September 18, 2017.

Mann, Randy. "Snowfall could hit local Top 10 list." *Coeur d'Alene Press*, January 9, 2020.

Mann, Randy. "Northwest summers get hotter and hotter." *Coeur d'Alene Press,* March 30, 2020.

Mann, Randy. "Randy Mann weather: Summer records could fall." *Coeur d'Alene Press*, July 19, 2021.

Mann, Randy. "Randy Mann: Early autumn might be in the air." *Coeur d'Alene Press*, August 23, 2021. The mean and hottest temperatures are from a September 28, 2021 email from Mann.

Mann, Randy. "Hot and dry now, but wait 'til fall." *Coeur d'Alene Press*, July 12, 2021.

Mann, Randy. "Tropical storms and hurricanes may be increasing very soon." *Coeur d'Alene Press*, August 29, 2022. Mann indicated this information is preliminary, since some data points in the 1980s are questionable and the 1990s had yet to be included in his analysis.

Marlon, Jennifer, Emily Goddard, Peter Howe, Matto Mildenberger, Martial Jefferson, Eric Fine, and Anthony Leiserowitz. "Yale Climate Opinion Maps." *Yale Program on Climate Change Communication*, January 23, 2024.

Marshall, Adrienne M., John T. Abatzoglou, Timothy E. Link, and Christopher J. Tennant. "Projected Changes in Interannual Variability of Peak Snowpack Amount and Timing in the Western United States." *Geophysical Research Letters* 46, no. 15 (2019): 8882-8892.

Martin, Shannon. "The impact of natural disasters on insurance rates in 2023." *Bankrate*, June 30, 2023.

Maté, Gabor, MD and Daniel Maté. *The Myth of Normal: Trauma, Illness & Healing in a Toxic Culture* (New York: Avery, 2022).

McCabe, Liam. "Can Heat Pumps Actually Work in Cold Climates?" *Consumer Reports*, August 2, 2022.

McCusker, Kelly and Hannah Hess. "America's Shrinking Ski Season." *Climate Impact Lab*. 8 (2018).

McCusker and Hess. (2018). Figure 3.

REFERENCES

McDonnell, Tim. "Why cutting your personal carbon footprint matters." *Semafor Net Zero,* April 21, 2023.

McDonnel, Tim. "Why Biden is spending $6 billion on green concrete and whiskey." *Semafor Net Zero,* March 27, 2024.

McIntosh, Bruce A., James R. Sedell, Russell Thurow, Sharon E. Clarke, and Gwynn L. Chandler. "Historical Changes in Pool Habitats in the Columbia River Basin." Abstract. *Ecological Applications* 10, no. 5 (2000): 1478-1496.

McKibben, Bill. "Global Warming's Terrifying New Math." *Rolling Stone*, July 19, 2012.

McKibben, Bill. "The Persistence of Fossil Fuels." Greta Thunberg, ed., *The Climate Book: The Facts and the Solutions* (New York: Penguin Press, 2022). Part IV: Chapter 5.

McKibben, Bill. "A Massive Win, and What It Means." Bill McKibben from *The Crucial Years,* January 26, 2024. billmckibben@substack.com.

McKibben, Bill. "The most epic (and literal) gaslighting of all time." Bill McKibben from *The Crucial Years,* March 5, 2024. billmckibben@substack.com.

McLean, Mike. "Ice storm leaves N. Idaho black and blue." *Coeur d'Alene Press*, November 20, 1996.

Michelson, Joan. "Wave of 'Anti-ESG' Investing Legislation, New Study Found." *Forbes,* August 29, 2023.

Miller, Blair. "MT Supreme Court denies state's request to pause Held decision pending appeal." *The Daily Montanan.* Cited in *Great Falls Tribune*, January 17, 2024.

Millward-Hopkins, Joel, Julia K. Steinberger, Narasimha D. Rao, and Yannick Oswald. "Providing decent living with minimum energy: A global scenario." *Global Environmental Change* 65

(2020): 102168. "'26 Coldest Since '50." *Coeur d'Alene Press*, December 30, 1968.

Mitchell, Deborah, "Moving History Forward: Organized Sports in early Coeur d'Alene." *Coeur d'Alene Press*, January 22, 2022. Museum of North Idaho special to the *Press*.

Munich RE. "Record thunderstorm losses and deadly earthquakes: the natural disasters of 2023." January 9, 2024.

Naishadham, Suman. "Here's how cities in the West, including Las Vegas, have water amid drought." Associated Press., May 24, 2022.

NASA. A gallon of gas = 20 pounds of CO_2. https://climatekidsnasa.gov/review/carbon/gasoline.

Nasdaq. "Amount of Electric Vehicle Owners by State." 2024. https://www.nasdaq.com/articles/ amount-of-electric-vehicle-owners-by-state.

National Association of REALTORS' 2023 Community and Transportation Preferences Survey, "Survey: Americans Prefer Walkable Communities," September 2023, nar.realtor/ commercial/create/survey-americans-prefer-walkable-communities#.

National Oceanic and Atmospheric Administration, global science.

National Weather Service. "What Are Snow Ratios?" https://weather.gov/arx/why_snowratios.

Naughton, Keith. "Ford cuts price of F-150 plug-in pickup, set to resume shipments." *Bloomberg News*. Cited in *Spokesman-Review*, April 12, 2024.

"Net-Zero America report." Princeton University, October 2021. -jzGNG5aebQx0ihWp&ab_channel=AndlingerCenterforEnergyandtheEnvironment.

"New study helps solve a 30-year-old puzzle: how is climate change affecting El Nino and La Nina?" *The Conversation*,

May 18, 2023. https://theconversation.com/how-new-study-helps-solve-30-year-old-puzzle-how-is-climate-change-affecting-El-Nino-and-La-Nina? Cai, Wenju, Benjamin Ng, Tao Geng, Fan Jia, Lixin Wu, Huojian Wang, Yu Lin, et al. "Anthropogenic impacts on twentieth-century ENSO variability changes." *Nature Reviews Earth & Environment,* (2023): 1-12.

Newell, Peter, Michelle Twena, and Freddie Daley. "Scaling behaviour change for a 1.5-degree world: challenges and opportunities." *Global Sustainability* 4 (2021): e22.

Ninth Circuit Court of Appeals decision. Juliana v. U.S. climate lawsuit, January 17, 2020.

NOVA Now (host). "The case of hurricanes and climate change." [audio], September 20, 2021.

Ocko, Ilissa B., Tianyr Sun, Drew Shindell, Michael Oppenheimer, Alexander N. Hrista, Stephen W. Pacala, Denise L. Mauzerall, Yangyang Xn, and Steven P. Hamburg. "Acting rapidly to deploy readily available methane mitigation measures by sector can immediately slow global warming." *Environmental. Research. Letters* 16, no. 5 (2021): 054042.

OCT, 2023. Our Children's Trust online event, "Powering Youth Climate Justice," December 15, 2023.

Office of Energy Efficiency & Renewable Energy, 2021.

Oil Change International. "Planet Wreckers: How 20 Countries' Oil and Gas Extraction Plans Risk Locking in Climate Change." September 12, 2023.

Oliveria, D.F. "CFD Fights 3 More Tubbs Hill Fires." Huckleberries Online blog, August 18, 2015.

"Overview of Fires in Idaho and Associated Costs," Natural Reserve Interim Committee, presentation by State Forester David Groeschl, Idaho Dept. of Lands, Oct. 16, 2015.

Parrish, Jake. "A bit of moisture helps, but we're not out of the woods yet." *Coeur d'Alene Press,* August 31, 2015.

PBS Newshour. "Chasing Carbon Zero." *NOVA/PBS.* Narrated by PBS correspondent Miles O'Brien. Dr. Lott is Senior Research Scholar and Director of Research at Columbia University's Center on Global Energy Policy.

PBS. *Killer Floods.* 2023. Accessed August 29, 2023.

PBS Newshour. "Road Safety," interview with Pete Buttigieg, February 2, 2023.

PBS Newshour. "Conservatives fight back against environmental and socially conscious investments." August 29, 2023. Narrated by PBS Economics correspondent Paul Solman.

PBS *Newshour.* "Heavy Industry," interview with Rebecca Dell, ClimateWorks Foundation, March 29, 2024.

PBS News Weekend. "Why Americans are lonelier and its effects on our health." January 8, 2023.

Pearce, F. "Copenhagen: The era of climate stability is coming to an end." *The Guardian.* November 30, 2019. Pearce's *With Speed and Violence: Why Scientists Fear Tipping Points in Climate Change* (Boston: Beacon Press, 2007, is a compelling read on the lurking danger of climate tipping points.

Pearson, Kit. "More federal, state forest lands closed due to wildfires." *Coeur d'Alene Press,* August 28, 2015.

Penney, Veronica. "Where Wind and Solar Power Need to Grow for America to Meet Its Goals." *The New York Times*, August 28, 2021

Phares, Dennis. "How Batteries Can Save Us from A Carbon Disaster." TEDxReno, August 21, 2022.

Philip, Sjouke Y., Sarah F. Kew, Geert Jan Van Oldenborgh, Faron S. Anslow, Sonia I. Seneviratne, Robert Vautard. "Rapid attribution analysis of the extraordinary heatwave on the Pacific coast of the US and Canada in June 2021." *Earth Systems Dynamic Discussions* 2021 (2021): 1-34.

Pinson, A.O., E. Ritchie, and H.M. Conyers. "DoD exposure to climate change at home and abroad." *U.S. Army Corps of Engineers, 2021.*

Pistochini, Theresa, Oswald Michal Dichter, Subhrajt Chakraborty, Nelson Dichter, and Aref Aboud. "Greenhouse gas emission forecasts for electrification of space heating in residential homes in the U.S." *Energy Policy* 163 (2022): 112813.

Pontecorvo, Emily. "The Heat Pump Manufacturing Boom is About to Begin." *Heatmap*, November 11, 2023.

Post Opinions Staff. "How we can combat climate change." *The Washington Post*, January 2, 2019.

"Power Problems Continue." *Coeur d'Alene Press*, November 24, 1996.

"President Biden's Infrastructure Law." *White House*, 2022.

"Providing decent living with minimum energy: A global scenario." *Global Environmental Change* 65 (2020): 102168. In addition, see University of Oxford economist Kate Raworth's *Doughnut Economics: Seven Ways to Think Like a 21st-Century Economist* (New York: Random House, 2017). An economy is considered prosperous when all twelve social foundations are met without over shooting any of the nine ecological ceilings. Amsterdam is the first major city to downscale doughnut economics, while Costa Rica has become the first regenerative nation, according to Raworth's website.

"Public Transportation Plan Update: Final Report," August 2012.

Raj, Anita, and Lotus McDougal, L. "Sexual violence and rape in India." *Lancet* 383, no. 9920 (2014): 865.

Ramsay, Georgina and Askland, Hedda Haugen. "Displacement as Condition: A Refugee, A Farmer and the Teleology of Life." *Ethnos* 87, no. 3 (2022): 600-621. doi.10.1080/ 00141844.2020.1804971.

Rand, Joseph, Nick Manderlink, Will Gorman, Ryan Wiser, Joachim Reel, Julie Mulvaney Kemp, Seongeun Jeong, and Fritz Kahrl. "Characteristics of Power Plants Seeking Transmission Interconnection as of the End of 2023." *Lawrence Berkeley National Laboratory*, April 2024. Originally noted in *Semafor NetZero*, April 26, 2024.

Raworth, Kate. "Towards 1.5°C. Lifestyles." Greta Thunberg, ed., *The Climate Book: The Facts and the Solutions* (New York: Penguin Press, 2022). Part V: Chapter 3.

Redd, Stephen C. "Asthma in the United States: burden and current theories." *Environmental Health Perspectives* 110, no. suppl 4 (2002): 557-560.

Reed, Mary Lou. "Frozen in Time." *The Inlander*, December 30, 2015.

Refugees, U.N. (2023). *UNHCR Global Focus: Climate Action*. Retrieved from UNHCR.

republicEn.org. (2020, February 26). "Living With Wildfire In the Era of Climate Change." Webinar with Dr. Crystal Kolden.

Rich, Max. North Idaho syndromic surveillance data. Communicable Disease Surveillance Specialist, Division of Public Health, Idaho Department of Health & Welfare.

Robson, David. "The '3.5% rule': How a small minority can change the world," BBC, March 3, 2023.

Rockstrom, Johan, Joyeeta Gupta, Dahe Qin, Steven J. Lade, Jesse F. Abrams, Lauren Andersen, David J. Armstrong, McKay, et al. "Safe and just Earth system boundaries." *Nature* (2023): 1-19. See also Rockstrom, Johan and Owen Gaffney. *Breaking Boundaries: The Science of Our Planet*. (New York: DK Publishing, 2021).

REFERENCES

Rogers, Luke. "U.S. Population Grew 0.1% in 2021, Slowest Rate Since Founding of the Nation." *United States Census Bureau*, December 21, 2021.

Rush, Claire. "Heat wave to hit Pacific Northwest one year after deadly event." *AP Report for America*. Cited in *Coeur d'Alene Press*, July 25, 2022.

Russell, Betsy Z. "Little on climate change: 'It's a big deal.'" *The Idaho Press*, January 16, 2019. Sarofim, Marcus C., Shubhayu Saha, Michelle D. Hawkins, David M. Mills, Jeremy Hess, Radley Horton, Patrick Kinney, Joel Schwartz, and Alexis St. Juliana. "Temperature- Related death and illness." No. GSFC-E-DAA-TN31167. *U.S. Global Change Research Program*, 2016.

Schweitzer, Cheryl. "Sustainable jet fuel manufacturer breaks ground in Moses Lake." *Basin Business Journal*, August 29, 2023.

Seebauer, Sebastian. "Why early adopters engage in interpersonal diffusion: an empirical study on electric bicycles and electric scooters." *Transportation Project Research, Part A: Policy and Practice* 78 (2015): 146-160.

Selle, Jeff. "Fires Reported, snuffed on Tubbs." *Coeur d'Alene Press*, August 18, 2015.

Seltzer, Molly, A. "Big but affordable effort needed for America to reach net-zero emissions by 2050, Princeton study shows." *Princeton University*, Andlinger Center for Energy & the Environment, December 15, 2020.

"Shareholder revolts over climate crisis rock ExxonMobil and Chevron." *The Hill*, May 27, 2021.

Sharma, Sapna, Kevin Blagrave, John J. Magnuson, Catherine M. O'Reilly, Samantha Oliver, Ryan D. Batt, Madeline R. Magee et al. *et al*. "Widespread loss of lake ice around the Northern

Hemisphere in a warming world." *Nature Climate Change* 9, no. 3 (2019): *227- 231.* https://doi.org/10.1029/2020GL089608. Sharma notes the 25°F cutoff came from running a number of lakes and years in a model to determine the energy required to sufficiently cool the lake to prime it to freeze. Very small lakes, like Twin Lakes, may still freeze, she noted. Sharma 2023 email to author.

Showalter, Michael and Samuel A. Rasche. "Ninth Circuit Dismisses Kids' Climate Case (Again)", *The National Law Review,* May 3, 2024.

Siegfried, Erica Spanger-, Kristina Dahl, Astrid Caldas, Shana Udvardy. "The U.S. Military on the Front Lines of Rising Seas." *Union of Concerned Scientists,* July 27, 2016.

Sightline Institute, "Aging Solutions are Climate Solutions." Interview with *Climate Resilience for an Aging Nation* (Island Press: Washington, D.C., 2023) author Danielle Arigoni, April 18, 2024.

Singletary, Robert. "Coeur d'Alene Beautiful & Progressive: The Depression Years." *Museum of North Idaho Newsletter* 38, no. 4.

"Sled Race Better Than C. Chaplin." *Coeur d'Alene Evening Press,* February 5, 1916. Museum of North Idaho volunteer researcher and author of *Hydromania,* Stephen Shepperd, tracked down the details of Coeur d'Alene's sled dog races. Thanks to Steve's diligence and keen eye, he helped the author find additional photos for this book.

Smith, Brad. "Forest Service asked to take a deeper look at project near Hayden Lake," *Idaho Conservation League,* September 17, 2021.

Smith, Carly Parnitzke, and Jennifer J. Freyd. "Institutional betrayal." *American Psychologist,* 69, no. 6 (2014), pp. 575-587.

"Snow Blanket Piles High; Trees, Drifts Close Roads." *Coeur d'Alene Press,* January 13, 1969.

"Snow Removal Bill Jumps To $36,000." *Coeur d'Alene Press*, January 22, 1969.

Solar Energy Industries Association. Solar Industry Research Data, December 6, 2022.

"Sources of Greenhouse Gas Emissions." *U.S. Environmental Protection Agency*, 2021.

Spokane Regional Clean Air Agency website. A recap of the 2021 wildfire smoke season.

Springer, S. "Why is it so hard to stop buying stuff?", *Boston Globe*, May 18, 2017.

Springmann, Marco, H. Charles J. Godfray, Mike Rayner, and Peter Scarborough. "Analysis and valuation of the health and climate change cobenefits of dietary change. *Proceedings of the National Academy of Sciences* 113, no. 15 (2016): 4146-4151. www.pnas.org/cgi/doi/10.1073/pnas. Cited in Hawken, Paul, ed. *Drawdown: The Most Comprehensive Plan Ever Proposed to Reverse Global Warming.* New York: Penguin Books, 2017, 39.

Stephens, Tod. "Work begins to build better batteries." *Spokesman-Review,* November 30, 2023.

"Storm Closing Routes." *Coeur d'Alene Press,* January 31, 1969.

Subramonian, Samantha. "Engine No. 1: The little hedge fund that shook Big Oil." *Quartz,* May 28, 2021.

Sullender, Amanda. "Spokane awarded $9.2M grant to overhaul safety, install bike lanes." *Spokesman-Review,* November 12, 2023.

Swiss Re Institute. "The economics of climate change: no action not an option." April 27, 2021.

Sylvia, Tim. "Renewable energy accounted for 21% of U.S. electrical generation in 2021." *PV magazine,* February 28, 2021.

Table 2, EPA Air Data: Air Quality Data Collected Outdoor Monitors Across the U.S., 2021.

Tabuchi, Hiroko, Ephrat Livni and David Gelles. "SEC Approves New Climate Rules." *New York Times*, March 7, 2024. Cited in *Spokesman Review,* March 7, 2024.

Takemura, Alison E. "Heat pumps outsold gas furnaces again last year and the gap is growing." *Canary Media,* February 13, 2024.

Tankersley, Jim. "UN's climate summit embraces AI, but with some reservations." *New York Times.* Cited in *Spokesman-Review,* November 4, 2023.

"TerraPower selects Kemmerer, Wyoming as the preferred site for advanced reactor demonstration plant." *TerraPower,* November 16, 2021.

"The Inflation Reduction Act — What It Is and What It Means for EV Adoption." *Zero Emissions Transportation Association* staff, August 10, 2022.

"The Montreal Protocol evolves to fight climate change." *U.S. Secretary of State*, September 2022. https://www/state.gov/u-s-ra tification-of-the-kingali-amendment.

Tidball, Les. "Lake accident kills Murphy." *Coeur d'Alene Press*, January 13, 1986. In the early 1900's, dozens of people routinely ice skated off Tubbs Hill (see Figure 5) and in the 1930s, trucks took supplies across the frozen lake to miners and loggers. Ice was thicker and better quality then. The loss of reliably thick lake ice is one toll of human-driven climate change. Due to Fred Murphy's many unselfish contributions to the community, Coeur d'Alene's Memorial Day Weekend parade was named in his honor, from 1988 to 2004.

Tiernan, Colin. "About 600 Failed to Finish Superheated Coeur d'Alene Ironman." *Spokesman- Review,* June 29, 2021.

"Top 10 most beautiful lakes in the U.S." *Vacasa, LLC.* https://www.vacasa.com/discover-the-top-10-most-beautiful-lakes-in-the-U.S.

Tucci, Joe, Janise Mitchell, and Chris Goddard. "Children's fears, hopes and heroes: modern childhood in Australia." *Australian Childhood Foundation* (Melbourne, Australia), July 3, 2007.

"Type 2 Diabetes Causes and Risk Factors," *WebMD,* June 22, 2023.

"Unlocking the Queue and Grid-Enhancing Technologies." The Brattle Group, February 1, 2021. https://watt-transmission.org/unlocking-the-queue/.

Van Westen, René, Michael Kliphuis and Henk A. Dijkstra. "Physics-based early warning signal shows that AMOC is on tipping course." *Science Advances* 10(2024): 1189. Cited in *Earthweek,* January 18, 2024.

Vaughan, A. "A mysterious rise in methane levels is sparking global warming fears." *New Scientist,* May 18, 2021. https://www.newscientist.com/article/mg25033350-700-a-mysterious-rise-in-methane-levels-is-sparking-global-warming-fears/.

Vestal, Shawn. "Smoky Augusts may be the new norm." *Spokesman-Review,* August 11, 2019.

"'Victory for our planet': Royal Dutch Shell must cut emissions." *Christian Science Monitor,* May 27, 2021.

Vohra, Kam, Alma Vodonos, Joel Schwartz, Eloise A. Marais, Melissa P. Sulprizio, and Loretta J. Mickley, L. "Global mortality from outdoor fine particle pollution generated by fossil fuel combustion: Results from GEOS-Chem," *Environmental research* 195 (2021): 110754.

Wang, Monica L., Marie-Rachelle-Narciese, and Pearl A. McElfish. "Higher walkability associated with increased physical activity and lower obesity among United States adults." *Obesity* 31, no. 2 (2023): 553-564. https://doi.org/10.1002/oby.23634.

Ward-Caviness, Calvin K., Mahdieh Danesh Yazdi, Joshua Moyer, Anne M. Weaver, Wayne E. Cascio, Qian Di, Joel D. Schwartz, and David Diaz-Sanchez. "Long-Term Exposure to particulate air pollution is associated with 30-day readmissions and hospital visits among patients with heart failure." *Journal of the American Heart Association* 10, no. 10 (2021): e019430.

"Water Q & A: What causes fish kills?" U.S. Geological Survey, Water Science School, *U.S. Global Change Research Program*, 217-246. https://doi.org/10.7930/JOTX3C9H.

WELC, Washington Environmental Law Center email, February 15, 2024.

Welsch, Quinn. "Smoke On the Water." *Spokesman-Review*, September 13, 2022.

Wesseler, Sarah. "American society wasn't always car-centric. Our future doesn't have to be either." *Yale Climate Communications*, October 3, 2023. https://yaleclimateconnections.org/2023/10/american-society-wasnt-always-so-car-centric-our-future-doesnt-have-to-be-either.

Whalen, John. "Walmart's EV gambit." *Spokesman-Review*, April 7, 2023.

"What will cities feel like in 60 years?" *University of Maryland Center for Environmental Science.* Future Urban Climates interactive web application. Information based on "Contemporary climate analogs for 540 North American areas in the late 21st century," by Matt Fitzpatrick of the University of Maryland Center for Environmental Science and Robert Dunn of North Carolina State, published in *Nature Communications*, February 2019.

Wheeler, David and Dan Hammer. "The Economics of Population Policy for Carbon Emissions Reduction in Developing Countries." *Center for Global Development Working Paper* 229, 2010. To obtain the discussed "incalculable" results would require a $39 billion worldwide annual investment. Cited in *Drawdown, 81.*

"Why Riding a Bike Is Good for You." *UCLA*, May 9, 2022.

Wiles, Kayla. "The whitest paint is here — and it's the coolest. Literally." *Purdue University.* Purdue News Service, April 15, 2021.

Williams, James H., Ryan A. James, Ben Haley, Gabe Kwok, Jeremy Hargreaves, Jamil Farbes, and Margaret S. Tom. "Carbon-neutral pathways for the United States." *AGU advances* 2, no. 1 (2021): e2000AV000284.

Winthrop and Karas, "Invest," Education for All Global Monitoring Report: *Pricing the Right to Education: The Cost of Reaching New Targets by 2030 Policy Paper 18.* Paris: United Nations Educational, Scientific, and Cultural Organization, 2015.

Wobus, Cameron, Eric E. Small, Heather Hosterman, David Mills, Justin Stein, Matthew Rissing, Russell Jones, et al. "Projected climate change impacts on skiing and snowmobiling: A case study of the United States." *Global Environmental Change* 45 (2017): 1-14.

Wolf, David, and H. Allen Klaider. "Bloom and bust: Toxic algae's impact on nearby property Values." *Ecological Economics* 135 (2017): 209-221.

Wolf, Joshua, and Robert Salo. "Water, water, everywhere, nor any drop to drink: climate change delusion." *The Australian and New Zealand journal of psychiatry* 42, no. 4 (2008): 350.

Wood, Zoe. "World is shifting to a more plant-based diet, says Unilever chief." *The Guardian*, February 4, 2021.

Wuerthner, George. "Guest Opinion — Climate Change, Not Fuels, Driving Large Blazes." *Spokesman-Review.* George Wuerthner has published two books on fire ecology and has traveled extensively throughout the West to view how large wildfires burn.

Yale Climate Connection. "Research: Solar panel prices are dropping without sacrificing quality." October 18, 2023.

"Youthful Mushers Compete as Dog Derby is Revived." *Coeur d'Alene Press*, February 19, 1938.

Zhou, Amanda. "UW Report Offers Solutions to Prevent Deaths During Heat Waves." *Seattle Times.* Cited in *Spokesman-Review*, July 11, 2023. https://www.seattletimes.com/seattle-news/environment/uw-report-offers-solutions-to-help-prevent-deaths-during-heat-waves

Index

A

Air Quality Index (AQI), 4, 69, 99, 103-104, 106-108
allergies, 109-111
    pollen concentrations increase, pollen season longer, warming-related, 110
And Peace Shall Return story, 256
aquatic ecosystems, 4 Cs needed in, 124-125
    cold, only C not compliance-based, 125
asthma, 106-108. 111
Avista's efforts to minimize wildfire risk, 212
    Wildland Urban Interface and, 212

B

bicycle use, expanding, as climate solution
    biking once a day decreases transport carbon emissions 67%, 174
    carbon footprint of cycling up to 30 times lower than fossil fuel car, ten times lower than driving electric, 174
Bosley, Chris interview, 174-176

C

car dependency impacts, 129-131
    15-minute city and, 129
    60% regularly lonely, social isolation and, 131
    inactive lifestyles & obesity, 95% diabetes cases and, 131
    increased risk of early death and, 131
    increased risk of heart disease & dementia and, 131
    loss of home as social connection and, 129
    tailpipe exhaust, 200,000 premature deaths annually and, 131
carbon capture and storage, 239-240
Cardinal Michael Czerny, 2023 Gonzaga climate speech, 154
climate and weather, distinction, 134
climate-aware therapy
    Dr. Marcia Rorty and, 204-206
climate change, economic damage by 2050 three times worse than COVID-19, 162
climate change, local impacts
    Kootenai County bird species loss, projected, 128
    loss of, and earlier melting snow, 1893-2023, 92-93
    loss of winter lake ice, 81-87
climate change, worldwide impacts of
    El Ninos and La Ninas stronger, more frequent, 135
    loss of winter lake ice, 84-85
    world ocean currents have declined 15% since 1950, 134
climate change movement
    local, state, national and international groups, 259-260

power of movements, 221-222
    3.5% rule of non-violent action, 222
    25% tipping point of, 222
climate change, worldwide, magnitude of
    36 billion tons of CO2 into atmosphere every year, 138-139
    traps heat equivalent of 500,000 Hiroshima-sized atomic bombs, 139
climate system tipping point, past disruptions
    ice sheets collapsed during last ice age, sea levels rose 65 feet over 400 years, 140
    temperatures crashed 29°F within decade, 140
Complete Streets ordinance, Seltice Way project and, 175
Conceivable Future group, reproductive justice & fossil fuels, 167-169
Coeur d' Alene Climate Adaptation Project, 203-204
Coeur d'Alene Eskimos ice hockey team, 40-41
Coeur d'Alene's climate future
    high & low emissions scenarios to 2100, 76
Coeur d'Alene's climate history
    1910s and 1920s, 36-39
    1916-1937, 39-40
    1933 Flood, 41-42
        historian Singletary and, 41-42
    1950s and 1960s, 43-46
    1968-1969 winter
        Coeur d'Alene Press stories and, 46-47
        local stories and, 47-52
    1970s to the 1990s, 52-55
    Ice Storm of 1996, 61-62
    windstorm of January 2021, 62-63
    the year 2015, 65-70
    snow drought, heat, wildfires and smoke, 65-68
    2021 Heat Wave, 70-73
    2022 heat waves, 73
Coeur d'Alene Tribe and climate change
    climate program priorities & seasonal calendar, 34
community adaptation plans, 201-202
consumerism and climate change
    cutting consumerism and degrowth, 165-166
    grow what needs improving, scale down what's destructive, 165
    house size tripled past 50 years, with smaller families, 163-164
    household demand, 60% of greenhouse gas emissions, 163
    1.5°C lifestyle, 164
cooling centers, Coeur d'Alene project, 207-209
cost of billion-dollar U.S. disasters past 10 years
    $1.1 trillion, and climate change, 160
cut food waste
    Enviro Certified Program, local stores and, 192
    wasted food, emissions impact of, 191

D

Day, Tabitha
    aggravated asthma & wildfire smoke, 4, 106-107
Department of Defense and climate change
    former Navy secretary Mabus & military risk, 149
    national security threat, threat multiplier, 149
    Naval Station Norfolk's risks, 150-151
displacement as global condition, climate change and, 16-17

E

eating Western diet, climate price tag, 190
eating plant-based diet, climate, health, and economic benefits of, 189-191
economic impacts of local climate change
    cost of fighting wildfires in Idaho 2-6 times higher since 2012, 118-119
    overcrowding, bigger issue than climate change in North Idaho State Parks, 118
    Row Adventures, lost revenue due to wildfires, smoke, low water, 116-117
    tourism, Kootenai County's biggest employer, 117-118
    outdoors, second biggest visitor activity in North Idaho, 117

INDEX

electric vehicles (EVs), health & cost benefits of, 184-184
EV charging stations, Coeur d'Alene, 183
EV commitments by car companies
    GM, Ford, Toyota, Jaguar, Volvo, 180
    EV sales, U.S., 2022-first quarter 2024, 181
Environmental Protection Agency proposed mileage rules
    Biden administration loosening of, 182
Exxon Mobil CEO comments and renewables, 157-158

F

forests, climate change benefits of, 125-126
    Forest Service prohibits cutting old-growth forests, 2023, 217-218

G

gardening and fishing casualties, local climate change and, 57-58
gifts of nature, Coeur d'Alene area, 23-25, 76-77, 203, 249
global warming
    distinct from climate change, 138
    thickening pollution blanket as, 6
"Global Warming's Six Americas", 146-147
Governor Little, public comments on climate change, 2019, 154-155
Grandpa, 9-13
grid storage, expanding, as climate solution, 236-237

H

harmful algae blooms
    health advisories, Fernan Lake and, 90-91
Hayhoe, Katharine, 153, 249
    "creation care" and psychological distance, 153
heat and local violent crime impacts
    arrests spike during high heat in 2015, 2017, 2021, 120
        disorderly conduct arrests related to high heat, 121
heat and violent crime increase above 85°F, study, 120

heat pumps, 185-187
    Gradient heat pump, 186
heat waves, air conditioning limits in, 74
Held v. State of Montana
    first constitutional climate trial, 17-20
home
    expressions of, 7-8, 12, 23-25, 130-131, 251
    qualities of home being lost, 4-6, 26-29, 54-61, 64-70
    clean air, 4
    comfortable climate, 6
    unrestrained enjoyment of outdoor recreation, 5
hotter summer temperatures, local, 5-6, 64-65
    community events and, 119-120
hunting and fishing Impacts, local climate change
    Denny Webster and, 1978 to present, 55-58

I

Idaho Fish and Game, Wildlife Action Plan, 2022
    plant and animal impacts from climate change, 126-127
Idaho's clean energy future, Jacobsen study and, 220
Inflation Reduction Act, 242-243
Infrastructure Investment & Jobs Act, 242
    new transmission lines and 500,000 EV chargers by 2030, 242
Inglis, Bob, 152-153
    RepublicEn group and, 152

J

Juliana v. U.S. climate lawsuit, 227-230
    60 Minutes interview and, 227-229
    Our Children's Trust state, national & international climate cases, 227, 230

K

keeping 80% of fossil fuels in the ground, 223-225
    Bill McKibben article and, 223-224
    Carbon Tracker Initiative and, 223

Kigali Amendment, worldwide phasedown of HFC gases and, 244-245

L

Liquified Natural Gas, U.S. planned export threat, 254
    Biden action and, 254
local global warming diagnosis, symptoms, treatment and prognosis, 249-250

M

McLain (Hughsby), Sarah
    memories and recent climate observations, 36-37, 50-51, 60-61
methane, U.S. emissions, 80% reduction, EPA rule, 244
methane, worldwide emissions, half warming since 1850s and, 245
    emissions reductions and, 246
    Global Methane Pledge, 246
minimize air conditioner use as climate solution, 195-196
Montana, climate change impacts and fossil-fuel- based energy system, 219
    Montana's clean energy resource, Prof. Jacobson and, 219-220
Montecito, California's wildfire success story, 213-215
    lessons for Coeur d'Alene, HAHR communities, 216-217
    using fire as a tool, 217
Murphy, Fred, 2-3

N

net-zero emissions in the U.S. by 2050, 237-238
Net-Zero America Report and, 238
Carbon Neutral Pathways report, 237
nuclear power as solution, 240-241
    next generation reactor in Wyoming, 241
    ten times as expensive as solar & wind, 241

O

Oil company shakeups, and climate change, May 2021, 156-157

P

physical health impacts of climate change, local, 100-110
    climate change as AMA health emergency, 2019, 102
    Kootenai Health ED visits, asthma, heat-related illnesses, 108-109
    Kootenai Health ED visits, respiratory illnesses, 108
    Marimn Health, asthma, COPD, heat-related visits, 107-108
    wildfire smoke, PM-2.5 impacts decreased lung function, asthma, allergies, heart attacks, strokes, 100-102
    Spokane-Coeur d'Alene area 14th most polluted air in U.S., 2019-2021, 101
    unhealthy air quality days, Spokane, 102-104
place irreversibly damaged, 113
Plant Based Treaty, 218
plant more trees, 190
Pope Francis climate encyclical, 2015, 154
pricing carbon & Citizens Climate Lobby, 225-226
Project Drawdown and educating girls as climate solution, 253-254
    incalculable return on investment, 254
psychological impacts of climate change, 111-116
    children 25% higher infant mortality rate in extreme heat, 113
    "climate change delusion." 114-115
    institutional betrayal and, 115-116
    suicide linked to drought, extreme heat, poor air quality, 111
    suicide rate, Idaho, fifth highest in U.S., 2020, 112
    violence, suicides, murders, burglaries increase, 114
    worsening bipolar mania, depression, schizophrenia, 114
    Lise Van Susteren, PhD, and children's impacts, 112-113

R

Reed, Mary Lou
    local winters in 50s and 60s, 45-46

INDEX

renewable energy, U.S., 184-185
    25% of U.S. electricity first half of 2023, 185
    solar panel prices dropped 99% past 40 years, 184
rising risks of wildfire, climate change, local
    Tubbs Hill fires, 2015, 210
    300% increase in future area wildfires, 97, 210

S

save water as climate solution
    Las Vegas, exemplary example, 194-195
    local basics to save water, 195
Security & Exchange Commission's rules, inform investors of greenhouse gas emissions, 2024, 158-159
Sila Nanotechnologies, Moses Lake plant, 233
Simpson, Rep. Mike (ID)
    comments on climate change, 2019, 155
ski resort, drop in ski season length, 2017 study, 94
    local impacts: Lookout, Schweitzer, 49 Degrees North, 94-95
    ski seasons, U.S., opening later, shorter, 94
Snow Water Equivalent, 93, 95
social influence, purchasing solar panels and, 166
    plant-based diets influence business decisions, 166
Solar Dominance Hypothesis, 233-234
solar photovoltaic system installations, Coeur d'Alene Tribe and, 231
solar power
    fastest-growing electricity source, 231
    hiring past oil & gas company employees into industry, 232
solastalgia, 30-31
sustainable aviation fuel, developing
    Twelve, Moses Lake plant, 232-233

T

Toderian, Brent, 171-172
    multimodal city, walking, biking & transit enjoyable, 172

top ten individual actions to slow global warming, 169-170
    five of top seven actions involve transportation, 170
toxic algae blooms, 5, 89-91, 118
Toyota breakthrough in EV battery technology, 160-161
transportation, biggest driver, greenhouse gas emissions, Idaho and U.S., 141-142

U

update transmission grid, 234-236

W

walkability and bikeability, expanding, as climate solution, 176-177
walkable city, benefits of, 177-178
    factors in, 197-198
walk more, as climate solution, 177-178
Walk Score, 171
walk safely, threats to, 178-179
    Idaho and Kootenai County, rising pedestrian fatality rates, high driving speeds, 178
Walmart EV commitment, 181-182
washing and drying clothes, recommended changes in, 187-188
what are we up against with climate change?, 141-142
    Idaho: 5,900 EVs sold vs. need 600,000 sold by 2032, 141-142
where to invest your money
    ESG investing and, 199-201
    fossil fuel industry returns, underperformance of, 201
whitest paint developed, Purdue University, 246-247
willful ignorance and climate change
    Galileo story, James Carse, PhD, example, 250-251

Y

Yale Climate Opinion Maps
    Kootenai County views vs. national average, 147-148
young people and climate change
    climate change as concern, 14-15, 17-19, 114-116
    climate change burden on, 21

LOSING HOME

www.ingramcontent.com/pod-product-compliance
Lightning Source LLC
Chambersburg PA
CBHW070736170426
43200CB00007B/543